Concise Chemical Thermodynamics

Third Edition

Concise Chemical Thermodynamics

Third Edition

A.P.H. PETERS

CRC Press
Taylor & Francis Group
Boca Raton London New York

CRC Press is an imprint of the
Taylor & Francis Group, an **informa** business

CRC Press
Taylor & Francis Group
6000 Broken Sound Parkway NW, Suite 300
Boca Raton, FL 33487-2742

© 2010 by Taylor and Francis Group, LLC
CRC Press is an imprint of Taylor & Francis Group, an Informa business

No claim to original U.S. Government works

International Standard Book Number: 978-1-4398-1332-4 (Paperback)

Library of Congress Cataloging-in-Publication Data

Peters, A. P. H.
    Concise chemical thermodynamics -- 3rd ed. / A.P.H. Peters.
        p. cm.
    Rev. ed. of: Concise chemical thermodynamics / J.R.W. Warn and A.P.H. Peters. 2nd ed. 2002.
    Includes bibliographical references and index.
    ISBN 978-1-4398-1332-4 (alk. paper)
    1. Thermodynamics. I. Warn, J. R. W. (John Richard William), 1935- Concise chemical thermodynamics. II. Title.

QD504.W38 2010
541'.369--dc22
                                                                        2010014708

Visit the Taylor & Francis Web site at
http://www.taylorandfrancis.com

and the CRC Press Web site at
http://www.crcpress.com

Contents

Preface

This is the third edition of a book that was first published by Dr. J. R. W. Warn in 1969. The book was successful for many years and in 1994 I took up the task of writing a revised second edition, which was published in 1996. Although the subject has not changed basically since the two most important publications of Gibbs in 1876 and 1878, I believe that it is necessary for a thermodynamic textbook to be up to date.

Today's students are comfortable with computers. As a result, I have written a number of new problems with the use of programs such as Mathcad or Excel in mind. Students should graph data, explore expressions and fit experimental data, and plot functions that describe physical behavior.

Also, for professionals and students who are taking graduate courses in thermodynamics, very good and user-friendly software are now available for their thermodynamic calculations. FactSage and HSC Chemistry are two of them. They were also used in the preparation of the third edition of this book.

All popular features of the previous editions are retained while new ones have been added. The main body of the text remains largely unchanged. I have added new material to several chapters. In Chapter 1, there is an extensive outlook on the world's current energy consumption and the role of renewable energy in the future. In Chapter 6, an example of an exothermic reaction is added. In the so-called Mond process for extracting and purifying nickel, the "battle" between enthalpy and entropy is explained as a function of temperature. In Chapter 7, it is shown how a plot of the Gibbs energy for a reaction mixture versus the extent of reaction is calculated with the help of Mathcad. Students are asked to do this themselves in some additional problems. In Chapter 9, the Lambda sensor, which reduces vehicle emissions, is explained electrochemically with the Nernst law. In Chapters 10 and 11, the production of silicon in an arc (oven) furnace is described and calculated in detail with the aid of the professional FactSage software.

After working many years with this book, I have seen students making the same mistakes over and over again in some problems. Therefore, these problems have been reworked and more hints are given. Furthermore, new problems have been added to several chapters.

The list of suggested readings has also been updated. A solutions manual that contains solutions to all exercises in the text is available as a PDF file for instructors only. I also used the opportunity to correct old typographical mistakes, and if readers should detect some in this new edition, please do not hesitate to contact me at: a.peters@hszuyd.nl.

I would like to thank two of my students: Marc Linssen, for preparing a usable Word file from the e-book version of the second edition, and especially Jurrie Noordijk, who did a great job by retyping all tables, preparing numerous figures, and retyping all formulas with MathType.

I am very glad to know Prof. Dr. Klaus Hack, who is the managing director of GTT Technologies (Gesellschaft für Technische Thermochemie und –physik mbH) and professor at the Rheinisch-Westfälische Technische Hochschule in Aachen, Germany. He helped me with the preparation of Chapters 10 and 11 by providing data from the Scientific Group Thermodata Europe database, and with the computations by the FactSage program. I also am grateful for the discussions via e-mail about things I was not sure of.

I also wish to thank Annti Rone and Satu Mansikka from Ouotec Research Oy company in Finland. They provided me with the latest HSC Chemistry software program for free. I used this program for numerous calculations in the third edition and I can recommend it to all instructors who teach and students who follow courses in thermodynamics. For providing me with a photo of a silicon arc furnace for the cover of the third edition, I would like to thank Dr. Ing. Rolf Degel, general sales manager of SMS Siemag AG in Düsseldorf, Germany. And last, but not least, I would like to thank CRC Press, which has contributed to this new edition.

Ir Toon Peters
Heerlen, The Netherlands

Preface to the Second Edition

I switched to higher professional education in 1989, having worked as a secondary school teacher for four years. I started to work as a chemistry teacher in the higher laboratory education of the Hogeschool Den Haag, where I was requested to teach thermodynamics. Since a suitable course book was not available, I was at liberty to introduce a book of my own choice on thermodynamics. This choice was easily made after comparing a number of books: The first edition of *Concise Chemical Thermodynamics* explained thermodynamics in a simple way without slipping into complicated mathematical deductions. In addition, the book stood out due to the many practical examples that were used to illustrate the possible applications of thermodynamics in many disciplines. Owing to a pending merger of the higher laboratory education of the Hogeschool Den Haag with another Hogeschool, I switched to the higher laboratory education of the Hogeschool Heerlen, now Hogeschool Limburg, in 1990. Here, again, I was asked to teach thermodynamics and was allowed to choose my own course book. As my experience with Dr. Warn's book was extremely positive, I naturally introduced his book at my new school.

However, I had noticed that the book had not been updated since the first edition. After having used the book for five years, I had drawn up a list of suggestions and improvements which I sent to the publisher in 1994. It appeared that the publisher had intended to have the book revised for some time and was looking for a suitable candidate for the job. When I was asked to do this, I consented. I have tried to bring the book up to date and, where necessary, to introduce improvements. Without the help of a number of people, I would not have been able to complete this task successfully. I would like to thank Prof. Dr. Ir Erik Cordfunke and Ir Robert de Boer of the Netherlands Energy Research Foundation at Petten, who provided data for Chapters 1 and 9. I am very grateful to Dr. Klaus Hack of GTT Technologies at Herzogenrath, Germany, who not only provided data for Chapter 11 and Appendix III but also allowed me to use the computer program MicroTherm. Furthermore, I would like to thank my colleague, Mrs. Sis Achten, whom I often bothered with English-language problems. Finally, I would like to thank one of my students, Sjef Cremers, for typing and checking all the data used in Appendix III. Last, but not least, I would like to thank everyone at Chapman & Hall who have contributed to this new edition.

Ir Toon Peters
Heerlen, The Netherlands

Preface to the First Edition

Thermodynamics, like classical music, is an acquired taste. The initiation must be sensitively carried out; otherwise, the mathematical rigor, like the formal structure of the music, acts to discourage a deeper relationship. It is sad but true that some students, both of thermodynamics and of Bach, never recover from the initial shock. In this, we are all losers. In this book, therefore, I have tried to present thermodynamics in a way which uses little mathematics, but which includes many practical and lively examples. I hope it will provide a basis for an introductory course at Honors Degree level, and will also suffice for Higher National Diploma and Certificate students.

In general, my aim has been to develop an understanding of Gibbs free energy fairly early on and then to apply this concept in several different fields. I have used many examples from chemical industry, in the firm belief that the basic *usefulness* of the subject must be demonstrated. I have therefore included a brief discussion of Ellingham diagrams in Chapter 10 and, for a similar reason, have omitted any mention of the Carnot cycle. I have used symbols consistent with the recommendations of the International Union of Pure and Applied Chemistry and have adopted units of the Système International d'Unités. Accordingly, all thermochemical data are given in joules, not calories. At the time of writing, general agreement has not been reached on the replacement of atmosphere and gram mole by bar and kilogram mole, and so the older units have been retained.

During the writing of this book, I have drawn on the experience and advice of many individuals and organizations. I am very grateful to Drs. John Golden and Ron Perrett for reading major parts of the manuscript, and for helpful comments and discussions. In the writing of Chapter 10, I was much indebted to the personnel of Magnesium Elektron Limited, British Hydrocarbon Chemicals Limited, and the British Iron and Steel Research Association (who provided data for Figure 10.1). Thanks are also due to the Dow Chemical Company (for permission to reproduce sections of the JANAF tables), National Bureau of Standards, Washington, D.C. (for Figure 3.1 and for data from which Table 2.1 was compiled), and Royal Institute of Chemistry (for permission to include questions from previous examination papers, which appear herein as Problems 7.5, 7.7, 8.5, 9.8, and 9.11). Miss Hazel Hawkes, who typed the manuscript against great odds, also has my gratitude. Notwithstanding all of this help, I would never have completed this task without the continuous and firm support of my wife and family; to them, I am indeed very grateful.

My publishers have worked long and well in the removal of errors and ambiguities from the manuscript. It is possible, however, there may be some which remain, which I have hidden too well for detection by even their practiced eyes. To me must fall the blame for these errors; I would, however, be glad to know those detected by readers.

J. R. W. Warn
Hertford

Author

I was born in 1955 in Maasbracht in the south of The Netherlands. From 1968 to 1973, I attended high school at the Bisschoppelijk College in Roermond. From 1973 to 1977, I studied chemistry at the HTS (College of Advanced Technology) in Eindhoven. After graduation from this college in 1977, I went to the University of Technology in Eindhoven to study chemical engineering. One of the principal subjects was thermodynamics. In my final year I did research on solid-state nuclear magnetic resonance.

I interrupted this study for one year to serve in the army, and in 1984 I graduated from this university. Besides chemistry I also had other interests. In 1973, I took up correspondence chess and in 1998 I was rewarded with the title of International Correspondence Chess Master.

During my student days, I came to be acquainted with the music of Frank Zappa, who is still one of my favorite musicians. In the 1980s and 1990s, I also participated in numerous marathon events worldwide. Boston marathon is my most favorite one.

After graduating from college in 1984, I worked for four years as a chemistry teacher at a secondary modern school. In 1989, I switched to higher vocational education. In the school year 1989–1990, I was lecturer in physical and analytical chemistry at the Hogeschool Den Haag in The Hague. Due to a reorganization, I left this school in 1990 and went to the Hogeschool Heerlen (now Zuyd University).

At Zuyd University, I am currently employed as a senior lecturer in physical and analytical chemistry and internship coordinator for the chemistry and chemical engineering department. From 2004 until 2009, I was also coordinator of the chemistry department and starting in 2010 I am coordinator internationalization for our faculty. At Hogeschool Den Haag and Zuyd University, I was asked to lecture on thermodynamics because this was not a popular subject among lecturers and students (probably because most textbooks are rather difficult to understand for undergraduate students) and because I liked thermodynamics.

At both schools, I was free to choose a new textbook. After comparing several books, I chose Dr. Warn's *Concise Chemical Thermodynamics*. With this book, I was very satisfied. After sending a lengthy list of suggested corrections to the publisher in 1994, I was asked to revise the book. In 1996, the second revised edition was published by Chapman & Hall and in 2010 the third edition was published by CRC Press.

A. P. H. Peters

Symbols and Abbreviations

The symbols and abbreviations given below are based on *Quantities, Units and Symbols in Physical Chemistry*, 2nd ed., published by the International Union of Pure and Applied Chemistry, Blackwell Scientific Publications (1993).

a	(relative) activity
(aq)	aqueous solution
atm	atmosphere, unit of pressure
bar	bar, unit of pressure
c	concentration
°C	degree Celsius
C_P	heat capacity at constant pressure
C_V	heat capacity at constant volume
e	elementary charge
E	electromotive force
F	Faraday constant
f	fugacity
G	Gibbs energy (F in some books by American authors)
(g)	gas phase
H	enthalpy (heat content)
i	electric current
k	Boltzmann constant
K	equilibrium constant
K	Kelvin
kg	kilogram, unit of mass
L	liter, unit of volume
(l)	liquid phase
log	decadic logarithm
ln	exponential logarithm
m	molality (of a solute)
M	molar, unit of concentration
N_A	Avogadro constant
N	amount of substance, chemical amount
P	pressure
p_B	partial pressure (the definition applies to entities B, which should always be indicated by a subscript or in parentheses)
Q	reaction quotient
q	heat added to system
R	gas constant (molar)

(s)	solid state
S	entropy
t	time
T	thermodynamic temperature
U	internal energy
V	volume
w	work done on system
χ	mole fraction, amount fraction
α	expansion coefficient
γ	activity coefficient
Δ	change of function of state, final less initial
μ	chemical potential
Σ	summation sign

OTHER SYMBOLS AND CONVENTIONS IN CHEMICAL THERMODYNAMICS

(i) Symbols used as subscripts to denote a chemical process or reaction

Ads	adsorption
At	atomization
c	combustion reaction
dil	dilution (of a solution)
dpl	displacement
f	formation reaction
fus	melting, fusion (solid → liquid)
imm	immersion
mix	mixing of fluids
r	reaction in general
sol	solution (of solute in solvent)
sub	sublimation (solid → gas)
trs	transition (between two phases)
vap	vaporization, evaporation (liquid → gas)

(ii) Recommended superscripts

‡	activated complex, transition state
E	excess quantity
id	ideal
∞	infinite solution
*	pure substance
0	standard

(iii) Examples of use of these symbols

The subscripts used to denote a chemical process, listed under (i) above, should be used as subscripts to the Δ symbol to denote the change in an extensive thermodynamic quantity associated with the process.

Examples

$\Delta_{vap}H = H(g) - H(l)$, for the enthalpy of vaporization
$\Delta_{vap}H_m$ for the molar enthalpy of vaporization
$\Delta_s^l H^0 = \Delta H^0(l) - \Delta H^0(s)$, the change in (molar) standard enthalpy when a substance changes from solid to liquid state.

The standard enthalpy of combustion of gaseous methane is $\Delta_c H^0$ (CH_4, g, 298.15 K) = -890.3 kJ mol^{-1}, implying the reaction:

$$CH_4(g) + 2O_2(g) \rightarrow CO_2(g) + 2H_2O(l)$$

The standard (internal) energy of atomization of liquid water is $\Delta_{at}U^0(H_2O, l)$ = 625 kJ mol^{-1}, implying the reaction:

$$H_2O(l) \rightarrow 2H(g) + O(g)$$

1 Energy

1.1 THE REALM OF THERMODYNAMICS

Our world is characterized by a multitude of natural phenomena. It is a world of change, of movement and energy–of storms, earthquakes, cosmic rays, and solar flares. The range and complexity of these changes is so great that it would seem the height of foolishness to attempt to find a common thread or theme that runs through all of them. Nonetheless, over the centuries, by patient and careful observation aided by the occasional flash of insight, men have come to a partial understanding of the factors involved. At the heart of such a science is the concept of *energy*. Thermodynamics, as we now know it, derived its name originally from studies of the "motive power of heat" and had to do primarily with steam engines and their efficient use. The main reason why the idea of energy was so poorly understood is that energy, the capacity for doing work, appears in so many different forms. However, after heat and work were understood to be but different forms of the same thing and calculations of the efficiencies of steam engines were carried out, thermodynamics was applied more generally to all changes, both chemical and physical.

Thermodynamics is a science of the *macroscopic* world. That is, it requires no prior understanding of atomic and molecular structure, and all its measurements are made on materials *en masse*. This is not to say that an understanding of molecular phenomena cannot help us to grasp some difficult concepts. The branch of the subject known as statistical thermodynamics has assisted greatly in our understanding of entropy, for example, but the basic theories of thermodynamics are formulated quite independently of it. This point is even more evident if we consider a complete description of, for example, the steam in a kettle of boiling water. A description in molecular terms would involve the position and nature of each particle, and its velocity at any instant. As there would be well over 10^{21} molecules present, this would be a humanly impossible task. On the macroscopic scale, however, we are glad to find that the chemical composition of steam, its temperature and pressure, for example, are quite sufficient to specify the situation. If we accept that such variables as temperature, pressure, and composition are a sufficient description of such a system, and if we are prepared to follow energy in all its various disguises, then we shall find that thermodynamics is a reliable pathfinder and guide to new and unexplored phenomena. We shall find that by taking relatively simple measurements such as heats of reaction and specific heats, we can predict the outcome, and even calculate the equilibrium constants, of changes that may never have been attempted before.

Thermodynamics is a reliable guide in industrial chemistry, plasma physics, space technology, and nuclear engineering, to name but a few applications. We shall now discuss some aspects of the two main fields of application, which spring from the first and second laws of thermodynamics. The first law is the energy conservation law, which requires a clear understanding of energy's disguises. The second law deals with the concept of entropy, an increase of which may be regarded as one of nature's two fundamental forces. (The other driving force is the minimization of energy.)

1.1.1 ENERGY BOOKKEEPING

The rate of physical and mental development of the species *Homo sapiens* was previously limited by naturally occurring genetic processes. Revolutionary extensions to man's faculties have been made in the last two centuries. They are the extension of his mental capacity that came with the introduction of the electronic computer in the mid-twentieth century, and the extension of his physique that came with the development of machinery in the Industrial Revolution. We shall be concerned with this second aspect. There is a close link between the per capita consumption of energy and the state of physical advancement of a nation.

The current (2009) economic downturn is dampening near-term world energy demand growth. In the International Energy Outlook 2009 (IEO2009) projections from the U.S. Energy Information Administration (EIA), total world consumption of marketed energy is projected to grow by 44% between 2006 and 2030 as economic recovery spurs future demand growth (Figure 1.1 and Table 1.1) [1].

The largest projected increase in energy demand is for the non-OECD economies. The OECD (Organization for Economic Cooperation and Development) groups 30 member countries in a forum to discuss, develop, and refine economic and social policy.

Quadrillion Btu

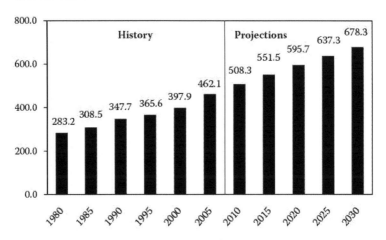

FIGURE 1.1 World marketed energy consumption 1980–2030. (Sources: History: Energy Information Administration (EIA), *International Energy Annual 2006* (June–December 2008), website www.eia.doe.gov/iea. Projections: EIA, World Energy Projections Plus, 2009.) A quadrillion BTU is equal to the amount of energy in 45 million tons of coal, or 1 trillion ft^3 of natural gas, or 170 million barrels of crude oil. In terms of electricity, 1 quad is equal to 293 GW h.

TABLE 1.1

World Marketed Energy Consumption by Country Grouping, 2005–2030 (Quadrillion Btu)

Region	2005	2010	2015	2020	2025	2030	Average Annual Percent Change, 2005–2030
OECD	241.3	242.8	252.4	261.3	269.5	278.2	0.6
North America	121.6	121.1	125.9	130.3	135.6	141.7	0.6
Europe	81.4	82.2	84.8	87.9	90.0	91.8	0.5
Asia	38.4	39.5	41.8	43.1	43.9	44.6	0.6
Non-OECD	220.7	265.4	299.1	334.4	367.8	400.1	2.4
Europe and Eurasia	50.6	54.0	57.6	60.3	62.0	63.3	0.9
Asia	109.4	139.2	163.2	190.3	215.4	239.6	3.2
Middle East	22.7	27.7	30.3	32.2	34.6	37.7	2.0
Africa	14.5	16.2	17.7	19.1	20.6	21.8	1.6
Central and South America	23.4	28.3	30.3	32.5	35.2	37.7	1.9
Total World	462.2	508.3	551.5	595.7	637.3	678.3	1.5

Sources: History: Energy Information Administration (EIA), *International Energy Annual 2006* (June–December 2008), website www.eia.doe.gov/iea. Projections: EIA, World Energy Projections Plus, 2009.

The OECD consists of like-minded countries, with the 30 member states all sharing a commitment to a market economy. The organization began in 1961 as a group of European and North American nations and has since expanded to include Japan, New Zealand, Australia, Mexico, Korea, and four former communist nations, the Czech Republic, Poland, Hungary, and the Slovak Republic.

Although high prices for oil and natural gas, which are expected to continue throughout the period, are likely to slow the growth of energy demand in the long term, world energy consumption is projected to continue increasing strongly as a result of robust economic growth and expanding populations in the world's developing countries. OECD member countries are, for the most part, more advanced energy consumers. Energy demand in the OECD economies is expected to grow slowly over the projection period, at an average annual rate of 0.6%, whereas energy consumption in the emerging economies of non-OECD countries is expected to expand by an average of 2.4% per year, as shown in Figure 1.2.

China and India are the fastest growing non-OECD economies, and they will be key world energy consumers in the future. Since 1990, energy consumption as a share of total world energy use has increased significantly in both countries. China and India together accounted for about 10% of the world's total energy consumption in 1990, but in 2006 their combined share was 19%. Strong economic growth in both countries continues over the projection period, with their combined energy use increasing nearly twofold and making up 28% of world energy consumption in

Quadrillion Btu

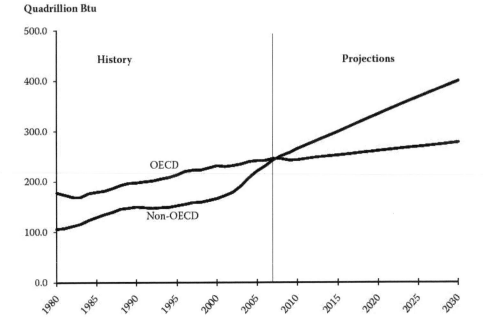

FIGURE 1.2 World marketed energy consumption: OECD and non-OECD, 1980–2030. (Sources: History: Energy Information Administration (EIA), *International Energy Annual 2006* (June–December 2008), website www.eia.doe.gov/iea. Projections: EIA, World Energy Projections Plus, 2009.)

2030 in the IEO2009 reference case. In contrast, the U.S. share of total world energy consumption falls from 21% in 2006 to about 17% in 2030.

Energy consumption in other non-OECD regions is also expected to grow strongly from 2005 to 2030, with increases of about 60% projected for the Middle East, Africa, and Central and South America. A smaller increase, about 36%, is expected for non-OECD Europe and Eurasia (including Russia and the other former Soviet Republics), as substantial gains in energy efficiency result from the replacement of inefficient Soviet-era capital stock and population growth rates decline.

It almost goes without saying that clean energy sources are preferable to energy sources that pollute the environment. Although we would rather limit ourselves to wind turbines and solar cells, for practical and economic reasons these are unable to meet our needs. Accordingly, current energy demand is principally met by fossil fuels such as oil, coal, and gas. How can we best meet this growing requirement?

Contributions from renewable energy sources cannot keep pace, while the prospects for a significant contribution from nuclear energy in many countries remain clouded by social factors. So, for the time being, we will continue to be dependent on fossil fuels, as can be seen in Figure 1.3.

This includes coal, since there are probably insufficient exploitable reserves of oil and gas to keep pace for long with increasing demand. No effort should be spared, therefore, to curtail the environmental pollution that accompanies the use of fossil fuels.

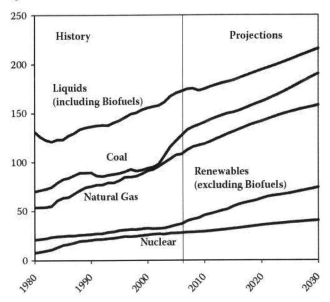

Quadrillion Btu

FIGURE 1.3 World marketed energy use by fuel type. (Sources: History: Energy Information Administration (EIA), *International Energy Annual 2006* (June–December 2008), website www.eia.doe.gov/iea. Projections: EIA, World Energy Projections Plus, 2009.)

It is essential that we should understand how far the laws of thermodynamics can help to clarify the various processes of energy conversion, or assist us in making efficient use of energy.

Consider some examples of energy conversion, both present and future.

(a) Our muscular energy (as a *mechanical,* or work-doing, form of energy) springs from the controlled combustion of carbohydrate foods. Some of the energy is used to warm us; some is used as mechanical energy. It also seems clear that a form of electrical energy is involved at an intermediate stage. Thus life itself depends on *energy conversion.*

(b) Plant life depends on converting radiant energy from the sun into the chemical energy of sugars and carbohydrates, which are photosynthesized from water and carbon dioxide.

(c) Solar energy is manifested in many different forms and, strictly speaking, in almost all forms of renewable energy, the sun is the primary source. Well-known examples of the use of solar energy are passive and active thermal solar energy, in which sunlight is used for space or water heating. A relatively new form of solar energy utilization is photovoltaic solar energy (PV). With PV (sun) light is converted directly into electricity using the PV effect. Since this process is fully "solid state," PV is clean, quiet, and reliable and can be used almost anywhere.

(d) Natural gas consists principally of methane, one of the gases that cause the greenhouse effect. Compared with other fossil fuels, natural gas is very clean. During combustion, however, atmospheric nitrogen reacts with oxygen to produce a mixture of nitrogen oxides. This reaction occurs because of an extremely high temperature of combustion and an excess of available oxygen. It is, therefore, a question of optimizing combustion techniques to minimize the effects of NO_x and other pollutants. As with coal gas, natural gas is a potential source material in the production of hydrogen. When it is heated with steam, a mixture of gases is obtained that contains hydrogen and CO_2.

The key to processes (a) through (d) is the abundant supply of energy. In the years to come, this requirement will be met from a variety of sources, and many different processes of energy conversion will be used. One thing, however, is certain. At all stages in the planning and execution of these schemes, basic thermodynamic concepts will form a central part. Such is the main theme of this book.

1.1.2 Nature's Driving Forces

In the 1880s, it was found that the combustion gases issuing from blast furnaces contained large amounts of carbon monoxide. The reactions in which monoxide is converted to dioxide, such as:

$$Fe_2O_3 + CO \rightarrow 2FeO + CO_2$$

or

$$FeO + CO \rightarrow Fe + CO_2$$

were not therefore occurring to completion. This seemed to indicate serious inefficiency in the furnaces. Perhaps insufficient time was being allowed for the reaction to be completed, or possibly these reactions were basically reversible, with small equilibrium constants for the forward reactions. If the smelters had known more about thermodynamics, they would have realized that the second solution was correct, and many firms would have been spared the expense of increasing the height of their blast furnaces. They did this in the vain hope that further opportunity would be given to the reactants to achieve equilibrium; in fact, the equilibrium could not be improved upon.

The tendency of a certain process to occur or not may be expressed in terms of the equilibrium constant. If we study the results of simple thermal measurements, thermodynamics will help us to calculate equilibrium constants of processes that have not been previously attempted, or which are difficult to study directly. It can answer the question "how far?" but is quite unable to answer "how fast?" A host of industrial processes, such as the cracking of light naphtha to form ethylene, or the synthesis of ethanol from ethylene and water vapor, depend initially on a very careful analysis of the equilibrium situation over a wide range of temperatures. Only when the feasibility of a new process has been demonstrated, by thermodynamic methods, is it necessary to build pilot plants and choose catalysts. Once a reaction strikes the rocks of an unfavorable equilibrium constant, no mere catalyst can salvage the wreck.

1.2 SETTING THE SCENE: BASIC IDEAS

Thermodynamics depends on a careful logical analysis of natural phenomena, and so we must first agree on a small number of basic terms.

1.2.1 SYSTEM AND SURROUNDINGS

The experiments we shall discuss will all be carried out in a small, well-defined space, usually in the laboratory, which we can study closely; this small region of space is called the *system*. The system may be simple or complex; it may be, for example, a container of chemicals ready to react, or a simple vessel of gas. There may well be facilities for stirring (to bring about homogeneity), or for measuring temperature or pressure. Sometimes the boundaries are only imagined, but they can always be represented on paper, so that everything outside the system is differentiated as the *surroundings*.

A system may undergo alteration from outside in two distinct ways. From time to time, either *chemical matter* or *energy*, in one of its forms, may move across the boundary. The energy may be mechanical energy transmitted by a shaft, electrical energy flowing through wires, or conducted or radiated thermal energy. The system is said to be *closed* if no matter is allowed to move across the boundary, and *adiabatic* if no thermal energy passes into or out of the system. If neither matter nor energy is transferred, it is said to be isolated.

1.2.2 FUNCTIONS OF STATE

When describing an experimental sample of, for example, two moles of methane, it is only necessary to specify a certain minimum number of variables in order to describe the system completely. Thus temperature and pressure would be quite adequate, although we could substitute volume or density for one of them. In fact, any two would automatically fix the remainder. For an ideal gas this follows directly from the general gas equation: $PV = nRT$.

These functions of state, or state variables as they are sometimes called, are closely dependent on one another, and describe the state or condition of the system as it is now, without reference to its immediate history. This lack of memory has an important consequence. If, for example, we are measuring the change of temperature of a tank of liquid heptane, from a "state 1" to a "state 2," the value of the temperature difference, $T_2 - T_1$, is quite independent of the intermediate steps of the change.

If we were considering the amount of thermal energy that had been added to achieve this rise in temperature from T_1 to T_2, we would find that it depended on the way in which it was accomplished. In one case, we might have raised the temperature by passing hot water through some heating coils. Alternatively, we might have used mechanical energy to agitate the liquid heptane; this too would result in a temperature rise, although no heat or thermal energy would have been added. Thermal energy and mechanical energy themselves would certainly not qualify as functions of state, but could alter with the whim of the experimenter and would depend on the route by which he chose to go from state 1 to state 2.

It should be pointed out here that a change of volume or of any other function of state is always measured as that of the *final state minus that of the initial state*. Mathematical abbreviations save time and space, and as a general rule we shall use Greek delta (Δ) to denote such a change, and Greek sigma (Σ) to denote "sum of." Thus on melting one mole of mercury, there is a volume change represented as follows:

$$Hg(s) \quad \rightarrow \quad Hg(l)$$

Molar volume (V_m)   14.134 mL   14.654 mL

$$\Delta V_m = \sum V_m \text{(products, final state)} - \sum V_m \text{(reactants, initial state)}$$

$$= 14.645 - 14.434$$

$$= +0.520 \text{ mL}$$

The positive sign implies an increase in volume, whereas a negative sign would imply a decrease. In quite general terms, for a change of a function of state in which products (final state) are formed from reactants (initial state):

$$\Delta\text{(function of state)} = \sum \left(\begin{array}{c} \text{function of state,} \\ \text{products} \end{array} \right) - \sum \left(\begin{array}{c} \text{function of state,} \\ \text{reactants} \end{array} \right) \quad (1.1)$$

Finally, it must be stressed that to measure a function of state implies that equilibrium exists within the system. We can generalize only about systems that are thermally, mechanically, and chemically in equilibrium. This condition is seldom met. Slight differences of temperature are often found even in a carefully controlled apparatus, and the experimental thermodynamicist must be constantly alert to these possibilities.

1.2.3 MECHANICAL WORK AND EXPANDING GASES

When we take the lid off a bottle of cola, the released carbon dioxide does work on the atmosphere. That is, it has to push back the air, which resists the expansion, to make room for itself. The same is true of steam from water boiling in a kettle and of many chemical processes. This is one of the many forms of energy that we must learn to recognize. The expression we shall now derive will be quoted often.

Consider the reaction of dilute hydrochloric acid on calcium carbonate to be taking place in a tall form beaker. The evolved CO_2 will push back the air. It is convenient to imagine a weightless piston, which just fits the inside of the beaker, to be moving up at the same time (Figure 1.4).

Mechanical work is equal to the product of the force, f, and the distance moved and, in this case, for a small movement dh, work $= f \, dh$. If A is the cross-sectional

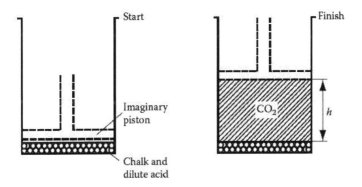

FIGURE 1.4   Evolution of carbon dioxide causes work to be done on gas in surrounding atmosphere.

area of the piston, and P is the resisting external pressure (usually atmospheric), the total work is equal to:

$$\int PA\,dh = \int P\,dV = P\Delta V. \tag{1.2}$$

If we assume that the gas behaves ideally, the general gas equation, $PV = nRT$, where n is the number of moles of gas and R is the universal gas constant, can be used to simplify the measurements.

In passing, it is as well to mention the numerical value of R. We know that a mole of gas at 1 atm occupies 22.413 L, and so, at 273.15 K, the equation of state becomes $1(22.413) = (1)R(273.15)$, which gives $R = 0.08205$ L atm mol^{-1} K^{-1}.

Alternatively, we may express R in J mol^{-1} K^{-1} (1 L atm $= 101.3$ joule), or in cal mol^{-1} K^{-1} (4.184 J $= 1$ cal, by definition). To summarize these calculations:

$$R = 0.082 \text{ L atm mol}^{-1} \text{ K}^{-1}$$

$$= 8.314 \text{ J mol}^{-1} \text{ K}^{-1}$$

$$= 1.9872 \text{ cal mol}^{-1} \text{ K.}^{-1}$$

From time to time, it will be necessary to use each one of these expressions.

1.2.4   THE ABSOLUTE TEMPERATURE SCALE

We all have a more or less well-developed sense of hot and cold, and understand the concept of temperature to mean that high values are associated with hot bodies and low with cold. The measurement of temperature on a numerical scale, however, has always presented problems. The basic difficulty is that the measured temperature depends on the thermometer being used.

The problem of the choice of thermometer is best understood by example. Let us begin by defining 100°C as the temperature of boiling water at 1 atm pressure, and 0°C as that of wet ice at 1 atm in contact with air. We can then calibrate a mercury-in-glass thermometer by using these two points. Assuming a smooth capillary, we can interpolate a value of 50°C exactly midway between the calibration points. We therefore define this intermediate temperature in terms of our chosen thermometer. However, if we construct thermometers using the e.m.f. of a thermocouple, or the volume of helium gas as the thermometric property, serious discrepancies arise. At 50°C, the thermocouple reads 49.27°C and the gas thermometer 49.97°C. Is there any reason for choosing one thermometer as a standard with which we can compare the others? Before answering this question, we must pay some attention to the properties of gases.

All gases expand on heating. Thus a liter of nitrogen gas at 0°C expands by approximately 0.00367 L when heated by 1°C. That is, if V_0 is the volume of the gas at 0°C, the volume at another temperature is given by:

$$V_T = V_0(1 + \alpha T)$$

where α is the coefficient of expansion, here equal to 0.00367, and T is the temperature in °C. The value of α is not absolutely constant, but varies slightly with the pressure of the gas (Problem 1.1). For an expansion from 0 to 1°C, we find that the ratio of volumes:

$$\frac{V_{(1°C)}}{V_{(0°C)}} = \frac{1/\alpha + 1}{1/\alpha}.$$

The term $1/\alpha$ has units of temperature. We are now able to define a new temperature scale such that $1/\alpha$ new degrees is equivalent to 0°C, and $(1/\alpha + 1)$ new degrees is equivalent to 1°C, and so on. This means that the gas volume is directly proportional

TABLE 1.2
Comparison between Thermometers, Calibrated Linearly from 0 to 100°C and the Perfect Gas Scale

Perfect Gas Scale (°C)	Expansion of Helium Gas at Constant Pressure	Expansion of Mercury in Jena Glass 59	e.m.f of Chromel-Alumel Thermocouple
400	399.95	412.6	400.0
300	299.98	304.4	297.8
200	199.994	200.8	198.3
100	100	100	100
50	50.001	50.03	49.3
0	0	0	0
–50	–50.002		–45.4

to temperature measured on our new scale. In fact, no gas obeys this relationship exactly, although helium comes very close to it. We nonetheless find that all gases obey it at the limit of zero pressure. The *hypothetical ideal gas* is thus defined as one that obeys it over all ranges. The temperature scale that results is named after Lord Kelvin (William Thomson), who was its originator.

The value of $1/\alpha$ for nitrogen at 1 atm is 272. Experiments with other gases indicate that the ice point, $0°C$, is equivalent to a value of near 273 K. The Kelvin scale is now defined with high accuracy such that the triple point of water (where ice, water, and water vapor are all in equilibrium, at $0.01°C$) has the temperature $273.1600°$ on the Kelvin scale. The triple point is more accurately defined than the ice point. On this basis, $0°C$ is 273.15 K. Measurement on this ideal gas scale is best conducted with a constant pressure helium thermometer, although there are small deviations from the absolute scale. A comparison of the four temperature scales discussed above is given in Table 1.2.

1.3 FORMS OF ENERGY AND THEIR INTERCONVERSION

We are today familiar with the idea that energy takes several forms. That is, we see in an electric motor, a dam, or an internal combustion engine, a common factor that we define as the capacity for doing work. The fact that the forms of the energy, electrical, gravitational, and chemical, are fundamentally different does not detract from their working usefulness. Such understanding is of fairly recent origin, however. The idea that heat is a form of energy, manifesting itself as "energy in transit" dates from 1842, when J. R. Mayer first published his hypothesis that heat and work were different forms of the same thing. On a voyage to the tropics, when bleeding a sick crew member, he noticed that the venous blood, usually dark and sluggish, was rich with oxygen and bright red in color. This was because less oxygen was needed for oxidation of food to maintain the body temperature than in colder parts of the world. He then considered the possibility that the combustion of a given amount of food led to both heating and working of the body, but in different proportions depending on the need for each. The nature of heat, the caloric fluid, was an enigma up to that time, and even now it is a difficult concept. What Mayer came to realize, to use our terminology, was that *energy is conserved*, and that the primary chemical energy is converted into heat or work as the need demands, with no energy being lost in the process.

Since that time, we have become aware of new disguises for energy. Sometimes the principle of the conservation of energy, which is the simplest expression of the *first law of thermodynamics* has seemed to be invalidated, but, in each case, the faith of those scientists who continued to believe in this great conservation law was vindicated.

A simple analogy will help to make the point. This analogy is based on a story of Nobel Prize winner in Physics Professor Richard Feynman (1918–1988), of the California Institute of Technology, with his permission.

A young child had many toys, but among his favorites was a set of sixteen building bricks, with which he played almost every day. Each evening, his mother would start the job of collecting them and tidying them away for the next day. Sometimes they would be hidden, but the mother remembered how many there were, and she believed in the "conservation of bricks" law, and so searched until they were found. One day it seemed that the child had put some bricks into his private lock-up box; it rattled. Although she could not see the bricks, she satisfied herself that there were, in fact, three bricks inside by weighing the box, knowing its empty weight, and calculating that:

$$\frac{(W_{(box\ plus\ bricks)} - W_{(box)})}{W_{(one\ brick)}} = \text{Number of bricks.}$$

On another occasion, the mother noticed that not only was the level of the goldfish pond a little higher than normal, but also that the fish were showing visible signs of consternation. By making calculations based on the apparent change of volume caused by the bricks, the law of conservation of bricks was verified once more. In each case, her awareness of the bricks took a different form, even though the total number was unchanged.

So it is with energy. It wears different disguises, but each may be penetrated by appropriate mathematical or experimental study. As an example, we may mention the "lost bricks" of radioactivity. The radioisotope of cerium, ^{144}Ce, for example, with a half-life of just more than 40 weeks, gives out atomic fragments and gamma rays at the rate of 25 W g^{-1}. Where does the energy come from? This and many similar nuclear processes are explained in terms of change of mass, according to the Einstein equation, which relates energy to mass and the velocity of light, by $E = mc^2$. Another example concerned certain beta-decay processes, in which a true energy balance seemed not to be achieved. This was due to the emission of a further particle, the *neutrino* (christened thus, the "little neutral one," by Enrico Fermi), which with its very small mass, and zero charge, was very difficult to detect.

Chemical energy is energy that is either stored or can be released through a distinct chemical reaction [2]. As such, it appears in many forms (Table 1.3), always in the form of some kind of potential energy. When a chemical reaction occurs, the energy of the chemical species may change and energy can be released to, or

TABLE 1.3
Types of Chemical Energy

Type	Examples
Chemical bonding	Connection between atoms
Electronic energy levels	Molecules after absorption of visible ultraviolet radiation
Vibrational/rotational energy levels	Thermally excited molecules or after absorption of infrared or microwave radiation
Osmotic energy	Membrane separated concentration differences[a]
Electrochemical energy	Elements capable of transferring electrons

[a] Especially interesting in the context of biological chemistry.

absorbed from, the surroundings. This can involve the exchange of chemical energy with another kind of energy or with another chemical system.

A chemical reaction, defined as a change in the chemical state of the system, is typically accompanied by a release of energy to, or an uptake of energy from, the surroundings. This energy transfer typically involves another form of energy, in patterns shown in Figure 1.5.

In many reactions that are permitted to proceed spontaneously, the energy emitted becomes thermal energy in the system and, eventually, its surroundings. This is what happens in the combustion of a fuel, such as coal, used to heat water for use in a turbine or engine. The efficient harnessing of chemical energy as thermal energy is, therefore, a critical part of the design of a combustion system.

Let us consider the combustion of coal, which involves the oxidation of carbon to carbon dioxide:

$$C + O_2 \rightarrow CO_2$$

The heat liberated in the combustion reaction is used to heat water vapor to produce high-temperature steam, which is then led into a turbine that, in turn, drives a magnet in a generator. There are, then, three energy conversions steps: (1) from

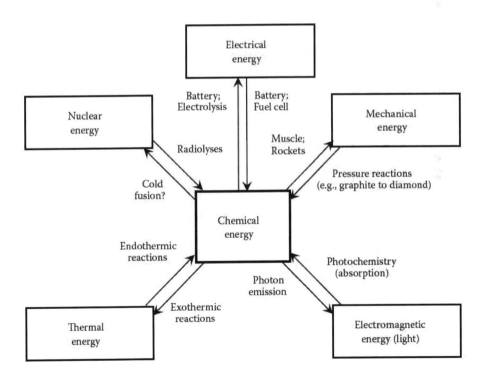

FIGURE 1.5 Relationship between chemical energy and other forms of energy, with examples. Note that a reaction can have energy transferred to or from several types of energy, as in the conversion of chemical energy into electrical energy and thermal energy.

chemical to thermal, (2) from thermal to mechanical, and (3) from mechanical to electrical.

In the combustion reaction, the chemical system of coal plus oxygen is converted into another chemical system, carbon dioxide, plus heat. Ideally, all the heat is captured in the thermal energy of the steam. However, the exhaust gases from the burner are very hot and remove heat to the atmosphere. Heat is also diverted to the vaporization of any noncombustible material in the fuel, especially moisture, and to bring the air used in combustion to the correct temperature. Energy is also released into the environment in the form of radiation, mostly in the infrared, from the hot surfaces of the equipment. The conversion of about 80% of the available chemical energy into actual thermal energy in the steam is the best practically achievable result.

The next step is the conversion of the thermal energy of the steam to the mechanical energy of a turbine. The efficiency of this step is limited by absolute thermodynamic constraints as described classically through an ideal heat cycle such as the Carnot cycle.

Practical machines achieve efficiencies in the order of 45%. Finally, the actual electrical generation step captures the mechanical energy of the turbine with more than 95% efficiency. Overall, the three-step electricity generation process yields about 33% of the chemical energy in the starting coal and oxygen as electrical energy.

There is an ongoing search for improved efficiency of these plants and for ways of reducing their impact on the environment. Coal gasification in combination with gas purification could, however, prove to be a technologically sound successor to conventional pulverized coal plants. This involves the production, from coal, of a firing gas (namely coal gas), which consists principally of hydrogen and carbon monoxide.

If coal can be represented by the formula C_nH_m, the gasification reaction can be written as:

$$2C_nH_m + nO_2 \rightarrow 2nCO + mH_2.$$

This means that there is a shortage of oxygen to burn the coal completely to carbon dioxide and water.

Coal gas can be used, for example, in a power station as a fuel for the production of electricity. Here, the gas is burned in a gas turbine, the residual heat being used in a steam cycle as described above.

A demonstration power station, with a capacity of 250 MW, was built in the early 1990s in Buggenum, in the Dutch province of Limburg. Sulfur-containing pollutants and oxides of nitrogen are removed from the coal gas so that, when it is used as a firing gas, the only products are water, carbon dioxide, fly ash, and slag. The net efficiency at full load is about 43%.

The removal of carbon dioxide from the flue gases is an expensive business. That is why it is better to remove carbon monoxide (from which the carbon dioxide is created) from the coal gas before it is used. This can be done by using steam to convert the carbon monoxide into hydrogen and carbon dioxide (shift reaction), which are then separated. The CO_2 released by this reaction can be stored underground, either in exhausted natural gas fields or in sealed water reservoirs. In this way, coal gasification can be used in the future as an intermediate phase in the production of hydrogen from coal. The hydrogen can subsequently be used as a fuel. The fuel cell could, potentially, be a major electricity producer, using hydrogen that has been purified in this way.

The fuel cell is highly efficient at converting hydrogen into electricity, without the use of combustion as an intermediate phase [3]. This conversion is accompanied by considerably fewer harmful side effects than in conventional installations, since the fuel cell generates practically no harmful waste products, and no noise nuisance whatsoever.

Fuel cells, which became well known as the auxiliary power supplies for the early manned space missions, are batteries in which the chemical fuels are continuously replenished. For journeys of medium duration, they offer the best energy-to-weight ratio, the cell and fuel being considered together as a unit. For longer flights, a radio-isotope generator is preferable because the weight of fuel for the fuel cell becomes excessive as distance increases. Today's energy use, based on the conversion of fossil fuels, is not without risk in the long-term. Harmful side products are produced by the combustion of fossil fuels.

Sulfur dioxide (SO_2) in the atmosphere is responsible for the acidification of the environment. Nitrogen dioxide (NO_2) pollutes the air primarily as a result of road traffic. Long-term exposure to NO_2 at concentrations of ≥ 40 µg m^{-3} is dangerous to people's health. Carbon dioxide (CO_2) is responsible for global warming.

Global warming, security of supply, and local air quality are strong driving forces to change the present energy system. Given the huge global demand for energy, no single solution can be imagined to make this energy supply more efficient, less carbon intensive, and more sustainable (i.e., using renewable sources to a large extent). The supplies of fossil fuels are limited. Nonrenewable energy sources do, as their name suggests, run out. Apart from their impact on global warming, they are finite. Based on the data we have today, we can predict the moment they are actually exhausted. Putting a date on these energy sources underscores the world's need for true sustainable energy sources, as given in Table 1.4 [4].

TABLE 1.4
Depletion of Nonrenewable Energy Sources

Nonrenewable Energy Sources	Total World Reserves Jan. 1, 2009	World Usage per Second	Estimated Year of Exhaustion
Natural gas (in m³)	174,436,171,550,404	92,653	2068
Oil (in barrels)	1,206,780,968,626	986	2047
Coal (in metric tons)	841,086,192,000	203	2140
Uranium (in metric tons U-235)	18,096	0.0000042222017	2144

Source: www.energy.eu (accessed August, 2009). With permission.

The EU is working to reduce the effects of climate change and establish a common energy policy. By 2020, renewable energy should account for 20% of the EU's final energy consumption (8.5% in 2005). The supplies of fossil fuels are regionally concentrated (see Table 1.5) [5]. As can be seen from Table 1.5, the European Union is very dependent on energy import, which is associated with global political risks. It should be clear that the usage of fossil fuels should be reduced and that all efforts should be made to switch to renewable energy sources with no risks for the environment. Once a deposit of fossil fuels is depleted it cannot be replenished–a replacement deposit must be found instead.

However, we must be aware that the types of figures from Tables 1.4 and 1.5 are discussed frequently worldwide and can change any moment. For example, 50 years ago in Slochteren, The Netherlands, one of the biggest conventional natural gas fields in the world was discovered with an estimated total amount of 2800 billion m^3 gas. In 2009, there was 1345 billion m^3 left, 33 billion m^3 more than was estimated in 2008. Conventional gas is easy to explore because it is enclosed in porous sand layers. From new geological studies in 2009, it appears that there is a total amount of approximately 500,000 billion m^3 of nonconventional gas in The Netherlands that is enclosed in micropores or is chemically or physically bonded to clay layers or coal,

TABLE 1.5
Energy Reserves Ranked by Country

Country	Natural Gas in Reserves[a] in Trillion m^3	Country	Oil Reserves[b] in Billions of Barrels
1. Russian Federation	44.65	1. Saudi Arabia	264.2
2. Iran	27.80	2. Iran	138.4
3. Qatar	25.60	3. Iraq	115.0
4. Saudi Arabia	7.17	4. Kuwait	101.5
5. United Arab Emirates	6.09	5. United Arab Emirates	97.8
6. USA	5.98	11. USA	29.4

Country	Coal Proven Reserves[c] in Million Tons	Country	Uranium Proven Reserves[d] as Recoverable Resources of Uranium in Tons
1. USA	242,721	1. Australia	1,243,000
2. Russian Federation	157,010	2. Kazakhstan	817,000
3. China	114,500	3. Russian	546,000
4. Australia	76,600	4. South Africa	435,000
5. India	56,498	5. Canada	423,000
		6. USA	342,000

[a] *Source:* Primary official sources and third-party data from Cedigaz.

[b] *Source:* OPEC Secretariat, *World Oil, Oil & Gas Journal.*

[c] *Source:* World Energy Council.

[d] *Source:* OECD NEA and IAEA.

200 times more than conventional amounts of gas! The big question is, "How to get to it?" Whether this gas is extractable depends on the success of new technologies and economical factors. Even if we find more fossil fuels and they are extractable, the future belongs to renewable energy.

1.4 FORMS OF RENEWABLE ENERGY

Renewable energy sources are energy sources that are continually replenished. These include energy from water, wind, the sun, geothermal sources, and biomass sources. Hydropower supplies 54% of the world increase in renewable energy, wind provides 33%, as is pictured in Figure 1.6.

Renewables are the fastest-growing energy source in the IEO2009 reference case, but fossil fuels still provide more than 80% of marketed energy in 2030.

1.4.1 SOLAR ENERGY

The sun has produced energy for billions of years. Solar energy is the sun's rays (solar radiation) that reach the earth. Solar energy can be converted to thermal (or heat) energy and used to heat water for use in homes, buildings, or swimming pools, and spaces such as greenhouses, homes, and other buildings.

Solar energy can be converted to electricity in two ways.

Photovoltaic (PV devices) or "solar cells" change sunlight directly into electricity. PV or solar cells offer the ability to generate electricity in a clean, quiet, and

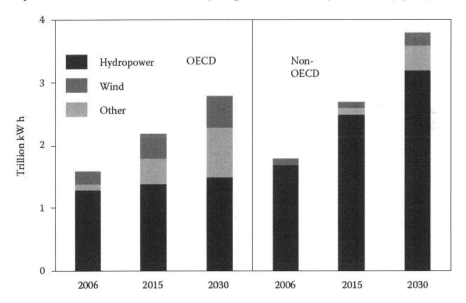

FIGURE 1.6 Hydropower supplies 54% of the world increase in renewable generation; wind supplies 33%. (Sources: History: Energy Information Administration (EIA), *International Energy Annual 2006* (June–December 2008), website www.eia.doe.gov/iea. Projections: EIA, World Energy Projections Plus, 2009.)

renewable way. Solar PV cells can be arranged in panels on a building's roof or walls, and can often directly feed electricity into the building. PV systems are often used in remote locations that are not connected to the electric grid. The variety of applications for solar electricity is numerous. PV cells are used in simple applications (e.g., calculators, watches, lighted road signs, and domestic and larger applications). Large PV systems can be integrated into buildings to generate electricity for export to the national grid.

Solar power plants indirectly generate electricity when the heat from solar thermal collectors is used to heat a fluid, which produces steam that is used to power a generator. Out of the 15 known solar electric generating units operating in the United States at the end of 2006, 10 of these are in California and 5 in Arizona. No statistics are being collected on solar plants that produce less than 1 megawatt of electricity, so there may be smaller solar plants in a number of other states.

1.4.2   WIND ENERGY

Wind is simple air in motion. It is caused by the uneven heating of the earth's surface by the sun. Since the earth's surface is made of very different types of land and water, it absorbs the sun's heat at different rates. During the day, the air above the land heats up more quickly than the air over water. The warm air over the land expands and rises, and the heavier, cooler air rushes in to take its place, creating winds. At night, the winds are reversed because the air cools more rapidly over land than over water. Wind turbines convert the kinetic energy from the wind into mechanical energy, which is then used to drive a generator that converts this energy into electricity. In the same way, the large atmospheric winds that circle the earth are created because the land near the earth's equator is heated more by the sun than the land near the North and South Poles. Today, wind energy is mainly used to generate electricity. Wind is called a renewable energy source because the wind will blow as long as the sun shines. This renewable source of energy has great potential in both onshore and offshore wind farms. Wind power is one of the cleanest and safest of all the renewable commercial methods of generating electricity.

1.4.3   HYDROELECTRIC POWER

We have been harnessing water to perform work for thousands of years. The Greeks used water wheels for grinding wheat into flour more than 2000 years ago. More recently, we have used the power of water to saw wood and power textile mills and manufacturing plants. Since the second half of the nineteenth century, the technology for using falling water to create hydroelectricity has existed. Of the renewable energy sources that generate electricity, hydropower is the most often used. In the United States, it accounted for 6% of electricity generation and 71% of generation from renewables in 2007. Because the source of hydropower is water, hydroelectric power plants must be located on a water source. Therefore, it wasn't until the technology to transmit electricity over long distances was developed that hydropower became widely used.

1.4.4 GEOTHERMAL ENERGY

Geothermal energy is heat from within the earth. We can use the steam and hot water produced inside the earth to heat buildings or generate electricity. Geothermal energy is a renewable energy source because the water is replenished by rainfall and the heat is continuously produced inside the earth.

1.4.5 BIOMASS ENERGY

The term biomass most often refers to organic matter such as timber and crops grown specifically to be burnt to generate heat and power. Biomass is organic material made from plants and animals. Biomass contains stored energy from the sun. Plants absorb the sun's energy in a process called photosynthesis. The chemical energy in plants gets passed on to animals and people that eat them. In order not to increase the amount of carbon dioxide in the atmosphere, it is important that wood burned as a fuel comes from sustainable sources. Biomass is sustainable and generally carbon neutral because the carbon released in the combustion process is offset by the carbon trapped in the organic matter by photosynthesis during its growth. This means that as trees are felled to be use as a fuel, more trees should be planted. That way, the carbon released during the combustion of the wood is reabsorbed by the new trees growing and the process is carbon neutral. Wood can be used as logs, wood chips, and wood pellets in wood/pellet burning stoves or wood chip/pellet boilers for space and water heating. Burning biomass is not the only way to release its energy. Biomass can be converted to other usable forms of energy such as methane gas or transportation fuels such as ethanol and biodiesel. Methane gas is the main ingredient of natural gas. Smelly stuff, such as rotting garbage, and agricultural and human waste, release methane gas–also called "landfill gas" or "biogas." Crops such as corn and sugar cane can be fermented to produce the transportation fuel, ethanol. Biodiesel, another transportation fuel, can be produced from leftover food products such as vegetable oils and animal fats. Biomass fuels provide about 3% of the energy used in the United States. People in the United States are trying to develop ways to burn more biomass and less fossil fuels. Using biomass for energy can cut back on waste and support agricultural products grown in the United States.

REFERENCES

1. History: Energy Information Administration (EIA). *International Energy Annual 2006* (June–December 2008). Available at: www.eia.doe.gov/iea. Projections: EIA. 2009. World Energy Projections Plus.
2. Wink, D. J. 1992. The conversion of chemical energy. *J Chem Ed* 69:108–111.
3. Publication Services ECN (Netherlands Energy Research Foundation). 1994. *Energy and Fuel Cells*.
4. www.energy.eu. Accessed August 2009.
5. Energy Information Administration, Office of Coal, Nuclear, Electric and Alternate Fuels. 2007. *The Role of Renewable Energy Consumption in the Nation's Energy Supply*.

PROBLEMS

1.1   Because the pressure of a gas increases with temperature, a ther-
      mometer can be made based on this behavior. The gas is placed in
      a constant-volume bulb (Figure P1.1) and its pressure is measured by
      the height h in the U-pipe. The gas in the bulb comes in thermal equi-
      librium with a sample, of which the temperature must be measured.
      The volume of the gas is kept constant by adjusting the position of the
      mercury reservoir so that the level in the left leg of the U-pipe is at the
      same height as temperature varies. The coefficient of expansion for
      nitrogen at 0°C varies slightly with pressure as follows:

Pressure (mm Hg)	α (K^{-1})
1000	$(3.674)\ 10^{-3}$
760	$(3.6708)\ 10^{-3}$
400	$(3.666)\ 10^{-3}$

      Determine from a graph of pressure versus $1/\alpha$ absolute zero in degrees
      Celsius as accurately as you can.

1.2   If 0.03 mol BaCO$_3$ reacts with dilute acid at 20°C, calculate the work
      done by the evolved CO$_2$ on the atmosphere ($R = 8.314$ J mol^{-1} K^{-1}).

1.3   Radioisotopes provide excellent low-power, long-life energy sources.
      The device SNAP-7B (System for Nuclear Auxiliary Power) powers
      an unattended lighthouse. It uses thermoelectric conversion; the hot
      junctions are clustered around the strontium-90 titanate source, which
      produces 0.45 W g^{-1}.

      (a) How many stages of energy conversion does this represent, and
          how does it compare with a conventional coal-burning power sta-
          tion with turbogenerators?
      (b) If 1 g of strontium-90 titanate is used for 5 years, to how many
          grams of fuel oil (calorific value = 43.93 kJ g^{-1}) is it equivalent
          (assume no diminution of power)?

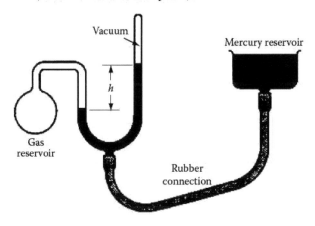

FIGURE P1.1   A constant-volume thermometer.

2 The First Law of Thermodynamics

In this chapter, a mathematical expression of the first law will be discussed. The two functions, internal energy, U, and enthalpy, H, will figure prominently. In addition, tabulations of standard enthalpy changes of formation, $\Delta_f H^0$, will be given for a number of compounds. From such tables it is possible to derive the amount of thermal energy associated with any reaction, as long as all reactants and products are listed.

2.1 STATEMENT OF THE FIRST LAW

All matter contains energy. It contains energy in respect of atomic and molecular motion, of interatomic forces, of magnetic and electric fields in the nucleus as well as outside it, of its physical state, and in many other ways. To calculate the total energy contained in a sample, we would first have to define a reference zero point. Such a zero would, in the end, be quite arbitrary. It is sufficient for us to consider only *changes* in the energy of a system, which, as we have seen, may be brought about principally by thermal or mechanical means. This and previous ideas are summarized in the formal statement of the first law:

$$\Delta U = U_2 - U_1 = q_{to} + w_{on}. \tag{2.1}$$

The term ΔU represents the change of internal energy of the system, q_{to} is the thermal energy (heat) *added* to the system, and w_{on} is the work *done on* the system. The subscripts "to" and "on" are emphasized here to avoid mistakes of sign in later calculations.

Internal energy U is a function of state. A change of internal energy from state 1 to state 2 must be considered only in terms of the initial and final states; *no route is specified*. The change from state 1 to state 2 may, however, be achieved in a multitude of different ways (Section 1.2.2). Heat may be added, or work may be done, or heat and work may both be involved in different proportions. That is, q and w both depend on the route followed during a particular process, and are not functions of state. These ideas may be expressed diagrammatically as in Figure 2.1. For a particular change, many routes are possible. To use an analogy, one can travel from London to Manchester by train, by air, or even (with perseverance) by sea, but the geometric distance between them is fixed at 260 km.

The work done on a system may be of many forms. For example, it might involve stretching of springs, lifting of weights, cutting of magnetic lines of force, or the

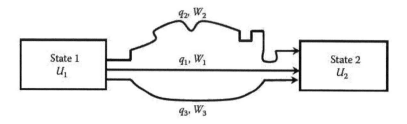

FIGURE 2.1 Internal energy is a function of state only, although many routes exist between states 1 and 2. Change in U is the same, whether it is $\Delta U = q_1 + w_1$, $\Delta U = q_2 + w_2$, or $\Delta U = q_3 + w_3$.

compression of gases. In the chemical applications that follow, we shall be concerned almost exclusively with gaseous pressure-volume work.

If we then consider an infinitesimal expansion of a gaseous system, dV, against an *external* pressure P_{ext}, the incremental work done on the system is, from Equation 1.2, $dw = -P_{ext}dV$. This becomes, for a significant amount of work, $w = -\int P_{ext}dV$.

The early experiments in space travel served to emphasize that the first law applies in orbit as well as on land. Temperatures in the first space capsules were much too high for comfort, reaching 40°C or higher for long periods. This was due partly to the effect of solar radiation and partly to the chemical and metabolic reactions occurring inside the vehicle. For the time being, we shall overlook the effect of the sun. The space vehicle can be considered a perfect example of a thermodynamic system; it has clearly defined boundaries and links with the surroundings that are easily spotted. The body's normal processes of carbohydrate oxidation and so on are exothermic, and if the capsule were to be completely isolated the temperature of the interior would rise indefinitely. That is, both q and w would be zero, and $\Delta U = q_{to} + w_{on} = 0$ (isolated). In effect, chemical energy is converted to thermal energy, but none is lost; U is constant.

Because the capsule is surrounded by an almost perfect vacuum, heat loss by conduction is negligible. At this point the situation looks desperate, but in fact significant radiation into space can occur even at 300 K. That is, for a steady temperature to be maintained, all surplus thermal energy must be radiated away. What about the radiation from the sun, which we have conveniently forgotten up to now? This energy, once received, must be reradiated as fast as it arrives. We can summarize the situation as we now see it. First, no work is done, so $w_{on} = 0$. Then,

$$q_{to} = q(\text{from sun}) - q(\text{radiated to space}) < 0,$$
$$\text{therefore}\quad \Delta U = q_{to} + 0 < 0.$$

That is, to maintain thermal equilibrium, *net* energy must leave the system. In fact, the capsule is carefully oriented so that a larger surface is radiating to deep space than is absorbing radiation from the sun. Also, the emissivities of the surfaces may be carefully chosen to give the best conditions.

2.1.1 REVERSIBLE EXPANSION OF AN IDEAL GAS

As the expansion of a gas proceeds at a constant temperature, its pressure drops. At any stage the amount of work it can do is determined by the opposing external pressure. If the opposing pressure is allowed to drop too quickly, less than the theoretical maximum work is done. If a railway engine develops wheel-slip, the work it does depends on the opposing frictional forces, not on its maximum (equilibrium) force. If a gas expands against an opposing pressure that is only infinitesimally less than its own, the term *reversible* is applied. (The two pressures are "virtually equal.") If the opposing pressure at any moment is increased infinitesimally, the direction of the process is reversed, and the gas is compressed again by a minute amount. Of course, such an overall process of compression or expansion would take an infinitely long time, and is never achieved in practice. It is nonetheless a useful theoretical model. In such a case, at constant temperature, the external pressure P_{ext} is *virtually equal* to P.

Thus, $w = -\int P dV$. For an ideal gas, $P = nRT / V$, and so:

$$w = -nRT \int \frac{dV}{V}$$

$$= -nRT \ln \frac{V_2}{V} \quad (T \text{ constant}). \tag{2.2}$$

2.1.2 CONSTANT-VOLUME PROCESSES

Many important experiments are carried out in closed containers called *bomb calorimeters* (Chapter 3). In this case, the volume remains constant, and so $dV = 0$, and $w = -\int P\, dV = 0$. Forms of work other than that due to gaseous expansion are already assumed to be zero. The first law may then be written:

$$\Delta U = q_v. \tag{2.3}$$

The subscript v implies constant-volume conditions. This means that the thermal energy added to a constant-volume system is equal to the increase in internal energy. This affords a simple and useful means of measurement.

Example 2.1

In a typical experiment, when 10 mmol (0.010 mol) of hydrocarbon were burnt in an excess of gaseous oxygen in a bomb calorimeter, 21.97 kJ of heat was evolved. Thus, $q_v = -2197$ kJ mol^{-1} for the process:

$$\text{Hydrocarbon} + \text{oxygen} \rightarrow x CO_2 + y H_2O,$$

and therefore $\Delta U = -2197$ kJ mol^{-1}. Note that, because heat is evolved, the internal energy decreases (ΔU has a negative sign). Such reactions are termed *exothermic*, whereas absorption of thermal energy at constant temperature characterizes an *endothermic* process.

2.1.3 CONSTANT-PRESSURE PROCESSES

The vast majority of chemical and other processes take place at constant pressure, usually of 1 atm. In these cases:

$$w = -\int P\,dV = -P\Delta V,$$

and

$$\Delta U = q_p - P\Delta V \quad \text{(only } PV \text{ work, } P \text{ constant).} \tag{2.4}$$

Here, q_p is heat added to the system during the constant-pressure process. It is not necessary to qualify ΔU in this way. (Why not?)

2.2 A NEW FUNCTION: ENTHALPY

We now define a new function, which is both important and supremely useful. This is the enthalpy of a system, which has the symbol H. An often-used alternative name is that of *heat content*, which is preferable insofar as it is partly self-descriptive. It can be confused with heat capacity, however, and so the term *enthalpy* will be used in this book. It is defined mathematically as:

$$H = U + PV. \tag{2.5}$$

Although it is difficult to visualize at this stage, it will become much more familiar with use. For the time being, we notice that, because both P and V (as well as U) are functions of state, H must itself be a function of state. That is, it describes the condition of a sample at a particular moment. Just as measurements of U in an absolute sense are quite inappropriate, so only changes in H are accessible to experiment.

Such an experiment or reaction would involve a "before" state, 1, and an "after" state, 2. We could write

$$H_2 = U_2 + P_2V_2$$

and

$$H_1 = U_1 + P_1V_1 \tag{2.6}$$

on subtracting

$$\Delta H = \Delta U + (P_2V_2 - P_1V_1).$$

If we allow all these functions to vary at once, the analysis is unnecessarily complex, especially when we realize that most processes occur in the open at 1 atm pressure. In these cases, $P_1 = P_2 = P$, and:

$$\Delta H = \Delta U + P \Delta V \quad (P \text{ constant}). \qquad (2.7)$$

If we rewrite this as $\Delta U = \Delta H - P\Delta V$ and compare it with Equation 2.4, we can identify q_p as ΔH; that is:

$$\Delta H = q_p \quad (\text{only } PV \text{ work, } P \text{ constant}). \qquad (2.8)$$

As in the case of ΔU, an exothermic change implies a negative sign for ΔH, an endothermic change implies a positive sign. Once again, it is stressed that q must always be qualified, because q_p is ΔH, although q_v is ΔU.

2.2.1 RELATIONSHIP BETWEEN ΔH AND ΔU

Determinations of the calorific value (CV) of many gaseous fuels, and of all solid fuels, are performed in constant-volume calorimeters. That is to say, values of $q_v = \Delta U$ are determined. When the fuel is burnt, however, at 1 atm pressure, additional energy of expansion (positive or negative) against the atmosphere is involved, and the value of q is actually realized; that is, $q_p = \Delta H$ may be significantly different. We start with Equation 2.7:

$$\Delta H = \Delta U + P\Delta V.$$

The change of volume $\Delta V =$ (volume of products – volume of reactants). Combustion involves the gases oxygen and carbon dioxide, and perhaps gaseous fuels, as well. If gases are involved in a reaction, they will account for by far the greatest part of the volume change, and we shall see that the volumes of solids and liquids are negligibly small in comparison. Suppose that we have n_1 moles of gas before the reaction, and n_2 moles of gas after it. Assuming the equation of state to be valid, we can write:

$$PV_2 = n_2 RT$$

and

$$PV_1 = n_1 RT$$

therefore

$$P\Delta V = \Delta n RT.$$

Substituting into Equation 2.7, we have:

$$\Delta H = \Delta U + \Delta n RT. \qquad (2.9)$$

This equation is very useful, as will be demonstrated in the following example.

Example 2.2

Consider the combustion of propane gas. The equation is:

$$C_3H_8(g) + 5O_2(g) \rightarrow 3CO_2(g) + 4H_2O(l).$$
$$n = 1 \qquad 5 \qquad\qquad 3$$

The letters g, l, and s in parentheses denote the phase of each chemical to be gas, liquid, or solid. Measurement of $q_v = \Delta U$ for this reaction, with the initial and final temperature near 25°C (298 K), the standard temperature for such measurements, gives $\Delta U = -2195$ kJ mol^{-1} of propane. Would more or less heat be liberated in a constant-pressure combustion unit giving identical products (this implies that product steam is condensed)? Six moles of gaseous reactants give only 3 mol gaseous product, so $\Delta n = (3 - 6) = -3$. That is, the atmosphere is *closing in on the system* while the process is occurring, and work is being done on the system. That is, additional energy is released (as heat) when constant pressure prevails. Using Equation 2.9, we find that:

$$\Delta H = \Delta U + \Delta nRT$$

$$= -2195 + \frac{(-3)(8.314)(298)}{1000}$$

$$= -2195 - 7.3 = -2202 \text{ kJ mol}^{-1}.$$

Thus, ΔH is, in fact, more negative than ΔU, although the difference is a very small proportion of the whole. [Moreover, the volume of water, 4(18) = 72 mL, is truly insignificant compared with 3(25) = 75 L, the approximate volume change of the gases.] For all that, the difference between ΔH and ΔU is frequently several percent when, for example, ΔH itself is small or the temperature is high.

2.3   USES AND CONVENTIONS OF ΔH

2.3.1   ENTHALPY CHANGE OF REACTION

During the building of the Aswan High Dam, many thousands of tons of concrete were poured. The primary setting process, that of hydration of the aluminosilicates, is exothermic. To avoid overheating and cracking, many miles of water-cooling pipes were embedded in the concrete. The reaction, and enthalpy change, may be represented as:

$$\text{Cement} + \text{water} \rightarrow \text{Concrete}, \Delta H = -28.9 \text{ kJ mol}^{-1} \text{ H}_2\text{O}.$$

That is, at constant pressure, 28.9 kJ of thermal energy is evolved per mole of water. The value of ΔH must be known before any reaction is carried out on a large scale, either industrially or in a laboratory. The heat must be removed (by heat exchanger or reflux condenser, as appropriate) because overheating could allow a reaction to get out of hand. Some Grignard reactions can occur dangerously fast if the temperature much exceeds 10°C.

Combustion processes constitute an important class of reactions, and the enthalpy changes associated with them give important information for operators of industrial plants as well as for internal combustion engineers. As an example, consider the hydrocarbon 3,3-dimethylpentane, C_7H_{16}, a component of high-octane gasoline. For combustion to carbon dioxide and liquid water:

$$C_7H_{16}(l) + 11O_2(g) \rightarrow 7CO_2(g) + 8H_2O(l),$$

ΔH is -4803 kJ mol^{-1}. Such a value would originate from an experiment with a constant-volume calorimeter, using the method described in Section 2.2.1 to convert ΔU to ΔH. The negative sign implies exothermicity.

Acid-base neutralization occurs exothermically with the formation of dissociated salt and water:

$$Na^+ + OH^- + H_3O^+ + Cl^- \rightarrow Na^+Cl^- + 2H_2O.$$

In the case of a strong acid and a strong base, not only the salt, but also the reactants are completely dissociated. The reaction is effectively that between hydroxonium and hydroxyl ions, and it will apply for *any* strong acid with *any* strong base. As a result, we find that the enthalpy change of neutralization is approximately constant for any such reaction, and equals -56.07 kJ mol^{-1}.

Not only chemical reactions, but *physical changes* may be assigned values of ΔH. Thus, a vaporization process, for example

$$Br_2 \, (l, \, 59°C, \, 1 \, atm) \rightarrow Br_2 \, (g, \, 59°C, \, 1 \, atm),$$

is assigned a value for ΔH of $+29.6$ kJ mol^{-1}. Thus, under these constant-pressure conditions, the enthalpy of bromine vapor is greater than that of the liquid by 29.6 kJ mol^{-1}. The alternative name of *heat content* for H is particularly easy to accept in this instance, as the vapor clearly contains "more heat." Not only vaporization, but fusion, sublimation, solid–solid phase changes, and also nuclear reactions may be described in terms of ΔH.

2.3.2 STANDARD ENTHALPIES OF FORMATION

It will be clear from what has been discussed that there are multitudes of possible reactions that might concern us and, therefore, multitudes of ΔH values. New compounds are being synthesized daily, and each will undergo many possible reactions. Clearly, it is almost impossible to determine ΔH for every possible change. Luckily, it is also unnecessary. It is sufficient, as we shall see, to know the standard enthalpy change of formation (sometimes abbreviated to *enthalpy of formation* or called *standard heat of formation*), for each compound with which we are dealing. The formation reaction is that in which the compound is formed from its elements in their standard states, that is, their most stable form under chosen standard conditions of temperature and pressure (stp). The standard pressure and temperature recommended by IUPAC since 1982 are $P^0 = 10^5$ Pa (= 1 bar) and 25°C (298.15 K). Up to 1982, the standard pressure was usually taken to be $P^0 = 101,325$ Pa (= 1 atm, called the *standard atmosphere*).

Under these conditions, for example, iodine is normally crystalline, hydrogen is gaseous, and sulfur exists as the rhombic crystalline form. In general, the compound formed will also be in its standard state. (This is not always the case, however. Thus it is possible, by a form of extrapolation, to calculate the standard enthalpy of formation of water vapor at 1 bar and 25°C, although it is normally liquid at this temperature. It is convenient to use such values for reactions at temperatures higher than, for instance, 100°C, because allowance will already have been made for the heat of vaporization of the water.) The omnibus expression for the enthalpy of formation is shown below.

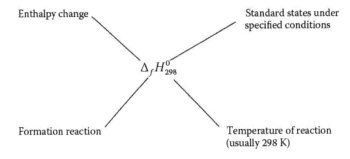

Enthalpy change

Standard states under specified conditions

$\Delta_f H^0_{298}$

Formation reaction

Temperature of reaction (usually 298 K)

Unless specified otherwise, $\Delta_f H^0_T$, where T is not 25°C, will refer to the reaction in which both reactants and products are in their standard states at the stated temperature, T. If aqueous solutions are involved, the standard state implies unit *effective* concentration, usually denoted aq. (This topic will be more fully discussed in Chapter 7.) Standard enthalpies of formation of salt solutions may be determined, and when $\Delta_f H^0$ for the hydrogen ion is taken *arbitrarily* to be zero, the values for individual ions may be quoted. In effect, we regard the hydrogen ion as an element. This is necessary because anions cannot be studied in isolation from cations.

Appendix III shows selected values of $\Delta_f H^0$ for a range of compounds, free atoms, and hydrated ions. Values are quoted in joules or kilojoules. However, values of a $\Delta_f H^0$ are only known to the nearest 50 J at best, and to 500 J or more in many cases. (Values of $\Delta_f G^0_{298}$ and of S^0_{298} are also shown, and these will be discussed in later chapters.) It should be noted that all values of $\Delta_f H^0$ for all elements are zero; this is a necessary consequence of the definition.

Example 2.3

The standard enthalpy change of formation for crystalline Na_2O is found to be -418.0 kJ mol⁻¹. This necessarily implies the following exact equation:

$$2Na(s) + \tfrac{1}{2}O_2(g) \rightarrow Na_2O(s), \Delta_f H^0_{298} = -418.0 \text{ kJ.}$$

Just more than 400 kJ are evolved per mole of crystalline sodium oxide formed. Sodium is normally solid and oxygen is normally gaseous under standard conditions, and so the starting materials and product are completely specified.

Example 2.4

The enthalpy of formation of crystalline $MgBr_2$ is:

$$\Delta_f H^0_{298}(MgBr_2(s)) = -526.0 \text{ kJ mol}^{-1}.$$

The temperature of 298 K implies the physical states shown in the equation:

$$Mg(s) + Br_2(l) \rightarrow MgBr_2(s).$$

However, it is quite possible that information is required at, for example, 373 K, at which temperature the vapor pressure of bromine is much more than 1 atm. Therefore, the standard state of bromine is now the gas, and the relevant equation becomes:

$$Mg(s) + Br_2(s) \rightarrow MgBr_2(s), \text{ with } \Delta_f H^0_{373} = -555.9 \text{ kJ mol}^{-1}.$$

Such a statement assumes the knowledge that bromine boils at 59°C, and the difference of approximately 29.9 kJ accounts for the heat of vaporization that is involved. Notice also that formation of the compound at 373 K requires the Mg(s) to be at 373 K, and that the heat of formation of Mg(s) at 373 K *is still zero* at this "new" temperature.

Example 2.5

The enthalpy of formation of barium ions is found to be −537.6 kJ mol⁻¹. We saw that these are to be formed from elements or hydrogen ions, and such a reaction is given by the complete equation:

$$Ba(s) + 2HCl(aq) \rightarrow H_2(g) + BaCl_2(aq),$$

which, because the ions are completely dissociated, may be written in the simplified form:

$$Ba(s) + 2H^+(aq) \rightarrow H_2(g) + Ba^{2+}(aq),$$

Apart from Ba^{2+} ions, every species is regarded as an element, with zero enthalpies of formation. It is therefore this equation to which the datum refers.

2.3.3 THE MANY USES OF $\Delta_f H^0$ DATA

We shall now see how the basic formation data may be used to provide information on a host of other reactions.

Two of the many possible chemical routes by which carbon may be converted to carbon dioxide are represented by the following equations:

Route I $C(\text{graphite}) + O_2(g) \rightarrow CO_2(g)$.

Route IIa $C(\text{graphite}) + \frac{1}{2}O_2(g) \rightarrow CO(g)$,

followed by

Route IIb $CO(g) + \frac{1}{2}O_2(g) \rightarrow CO_2(g)$.

It is clear that the initial state (reactants, carbon plus oxygen), and the final state (product, carbon dioxide gas) are identical for both routes. When enthalpy was introduced, it was agreed that it was a function of state, such that changes of it *did not depend on route*. In the present example, it follows that ΔH for Route I and that for Route II must be identical. That is to say that the overall change in H is equal to the sum of the values of ΔH for each step. Thus,

$$\Delta H(\text{I}) = \Delta H(\text{IIa}) + \Delta H(\text{IIb}).$$

Now Equation I represents formation of CO_2 and Equation IIa represents formation of CO. Values are:

$$\Delta_f H^0(CO_2(g)) = -395.5 \text{ kJ mol}^{-1}.$$

$$\Delta_f H^0(CO(g)) = -110.5 \text{ kJ mol}^{-1}.$$

Thus, $(-395.5) = (-110.5) + \Delta H(\text{IIb})$.

$$\Delta H(\text{IIb}) = -283.0 \text{ kJ mol}^{-1}.$$

That is, the combustion of CO to give gaseous CO_2 is accompanied by evolution of 283 kJ mol^{-1}. This, the principle of constant heat summation, often known as Hess's law, is thus seen to lead directly from the fact that H is a function of state. This idea is immensely powerful, because it enables ΔH^0_{298} values to be determined for any reaction, as long as the enthalpies of formation are known for each reactant and product.

Consider an equation of the general form:

$$A + B \cdots \rightarrow P + Q \cdots$$

for which we wish to know ΔH^0. Imagine this to take place in two stages, first the "unformation" of A, B, etc., into their elements, then the formation of P, Q, etc., from those same elements. That is, $A + B \cdots \xrightarrow{\Delta H^0_1}$ (elements in standard states) $\xrightarrow{\Delta H^0_2} P + Q \cdots$.

Now $\quad \Delta H_1^0 = -\left[\Delta_f H^0(A) + \Delta_f H^0(B) + \cdots\right] \quad$ (unformation)

and $\quad \Delta H_2^0 = \left[\Delta_f H^0(P) + \Delta_f H^0(Q) + \cdots\right] \quad$ (formation).

Thus, for the overall reaction:

$$\Delta H^0 = \Delta H_1^0 + \Delta H_2^0$$
$$= \left[\Delta_f H^0(P) + \Delta_f H^0(Q)\right] - \left[\Delta_f H^0(A) + \Delta_f H^0(B)\right].$$

This may be written in the abbreviated form:

$$\Delta H^0 = \sum\left[\Delta_f H^0(\text{products})\right] - \sum\left[\Delta_f H^0(\text{reactants})\right]. \qquad (2.10)$$

This general equation is seen to fit the prototype Equation 1.1, and will be used often.

Example 2.6

Fluorine has been considered for use as an oxidant in rocket propellants. Its high reactivity ensures high energy release, and consequently high thrust. Its reaction with ammonia:

$$NH_3(g) + \tfrac{3}{2}F_2(g) \rightarrow \tfrac{1}{2}N_2(g) + 3HF(g)$$

is being investigated. Let us calculate ΔH_{298}^0 for this reaction.

Ammonia is being decomposed and hydrogen fluoride is being produced. Fluorine and nitrogen are elements in their standard states, and thus have zero enthalpies of formation. By using data from Appendix III and Equation 2.10, we may write:

$$\Delta H_{298}^0 = \tfrac{1}{2}\Delta_f H^0(N_2) + 3\Delta_f H^0(HF) - \Delta_f H^0(NH_3) - \Delta_f H^0(F_2)$$

$$= 0 + 3(-273.3) - (-45.9) - 0$$

$$= -774.0 \text{ kJ mol}^{-1} \text{ NH}_3.$$

This represents a very high energy yield, especially when based on the mass of (fuel + oxidant). This energy is supplied by only $17 + \tfrac{3}{2}(38) = 74$ g of reactants (Problem 3.8).

Example 2.7

What heat is evolved on precipitation of one mole of $BaSO_4$ from solution of unit concentration?

Such a reaction could be represented as:

$$Ba^{2+}(aq) + SO_4^{2-}(aq) \rightarrow BaSO_4(s).$$

In this case, we find that:

$$\Delta H^0_{298} = \Delta_f H^0(BaSO_4) - \Delta_f H^0(Ba^{2+}) - \Delta_f H^0(SO_4^{2-})$$

$$= -1459 - (-537.6) - (-909.3)$$

$$= -12.1 \, kJ \, mol^{-1}.$$

Example 2.8

This example is taken from a paper by Margaret A. Frisch and J. L. Margrave [1] of the University of Wisconsin. They measured ΔH^0_{298} for the reaction:

$$NO(g) + CO(g) \rightarrow \tfrac{1}{2}N_2(g) + CO_2(g)$$

calorimetrically, and found a value of $-373.2 \, kJ \, mol^{-1}$ CO_2. It is necessary to use this result to check the value for the enthalpy of formation of nitric oxide. It is clear that, if values for $\Delta_f H^0$ are available for CO and CO_2, this can be done by using Equation 2.10. Thus,

$$H^0_{298} = \tfrac{1}{2}\Delta_f H^0(N_2) + \Delta_f H^0(CO_2) - \Delta_f H^0(NO) - \Delta_f H^0(CO)$$

$$-373.2 = 0 + (-395.5) - \Delta_f H^0(NO) - (-110.5)$$

whence $\Delta_f H^0(NO) = +90.21 \, kJ \, mol^{-1}$.

Notice that the formation reaction is endothermic.

An alternative, and equally useful form of Hess's law says, in effect, that if simple reaction equations are added (or subtracted) to give an overall reaction scheme, then so may the corresponding ΔH values be added (or subtracted) to give an overall value of enthalpy change. Two final examples will be given to demonstrate this alternative approach.

Example 2.9

The enthalpy changes for the following three reactions, which represent the formation of dilute boric acid solution, the dissolution of crystalline boric oxide, and the formation of liquid water, respectively, are as follows:

$$\text{I} \quad 2B(s) + 3H_2(g) + 3O_2(g) + aq \rightarrow 2H_3BO_3(\text{dil. solution})$$

$$\Delta H^0 = -2153 \text{ kJ}/2 \text{ mol } H_3BO_3,$$

$$\text{II} \quad B_2O_3(s) + 3H_2O(l) + aq \rightarrow 2H_3BO_3(\text{dil. solution})$$

$$\Delta H^0 = -17.4 \text{ kJ mol}^{-1},$$

$$\text{III} \quad H_2(g) + \tfrac{1}{2}O_2(g) \rightarrow H_2O(l)$$

$$\Delta H^0 = -285.8 \text{ kJ mol}^{-1}.$$

The term "aq" by itself represents an excess of solvent water. It is required to find ΔH for the reaction:

$$2B(s) + \tfrac{3}{2}O_2(g) \rightarrow B_2O_3(s) \qquad (2.11)$$

which represents the formation of boric oxide from its elements.

Problems such as these may be solved by combining equations in various ways, by observation, until an acceptable route for the required reaction is found. Here, one possible approach would be to observe that Equation 2.11 has as reactant $2B(s)$, and so does reaction I. The unwanted product of I, dilute boric acid solution, may be eliminated by performing II *in a reverse direction*; we note in passing that this involves an unwanted product of $3H_2O(l)$, produced chemically, and not as solvent. This may, in turn, be dealt with by performing III, in reverse, three times over. At this stage, we can write down the possible recipe I–II–3(III), which in full is:

$$2B(s) + 3H_2(g) + 3O_2(g) + aq \rightarrow 2H_3BO_3(\text{dil. solution})$$

$$2H_3BO_3(\text{dil. solution}) \rightarrow B_2O_3(s) + 3H_2O(l) + aq$$

$$3H_2O(l) \rightarrow 3H_2(g) + \tfrac{3}{2}O_2(g).$$

Notice that many items are both produced and consumed, and undergo no net change, and so may be deleted. On adding up these three reactions, we are presented with the desired process, Equation 2.11. (If we were not lucky the first time, we would learn to see which further reaction was needed as ingredient in the mixture.) The crux of the matter is this: Having followed the recipe I–II–3(III) for the equations, ΔH values may be treated in an identical manner. Thus,

$$\Delta H = \Delta H_I - \Delta H_{II} - 3(\Delta H_{III})$$

$$= -2153 - (-17.4) - 3(-285.8)$$

$$= -1278 \text{ kJ mol}^{-1} B_2O_3.$$

TABLE 2.1

Reaction Scheme for the Enthalpy of Formation of $Cs_2Te(s)$; (soln) Refers to $\{0.46 \text{ mol dm}^{-3} \text{ NaClO} + 0.5 \text{ mol dm}^{-3} \text{ NaOH}\}$; $\Delta H_{13} = -\Delta H_1 + \Delta H_2 + \Delta H_3 + \Delta H_4 + \Delta H_5 + \Delta H_6 - \Delta H_7 - \Delta H_8 + \Delta H_9 + \Delta H_{10} - \Delta H_{11} - \Delta H_{12}$

	ΔH (kJ mol^{-1})
1. $Cs_2Te(s) + \{4NaClO + 2NaOH\}(soln) = \{Na_2TeO_4 + 4NaCl + 2CsOH\}(soln)$	-1044.0 ± 1.5
2. $Te(s) + \{3NaClO + 2NaOH\}(soln) = \{Na_2TeO_4 + 3NaCl + H_2O\}$ (soln)	-652.73 ± 2.50
3. $Cs_2SO_4(s) + 2NaOH(soln) = \{2CsOH + Na_2SO_4\}(soln)$	$+12.08 \pm 0.02$
4. $2Cs(s) + S(s) + 2O_2(g) = Cs_2SO_4(s)$	-1443.36 ± 0.51
5. $Na_2SO_3(s) + NaClO(soln) = \{Na_2SO_4 + NaCl\}(soln)$	-352.48 ± 0.55
6. $2Na(s) + S(s) + 1\frac{1}{2}O_2(g) = Na_2SO_3(s)$	-1101.83 ± 0.72
7. $2Na_2SO_4(s) + (soln) = 2Na_2SO_4$ (soln)	-11.72 ± 0.06
8. $4Na(s) + 2S(s) + 4O_2(g) = 2Na_2SO_4(s)$	-2775.8 ± 0.8
9. $2\{NaOH \cdot 48.39H_2O\}(l) + (soln) = \{2NaOH + 96.78H_2O\}(soln)$	0
10. $2Na(s) + H_2(g) + O_2(g) + 96.78H_2O(l) = 2\{NaOH \cdot 48.39H_2O\}(l)$	-940.12 ± 0.16
11. $H_2(g) + \frac{1}{2}O_2(g) = H_2O(l)$	-285.830 ± 0.042
12. $95.78H_2O(l) + (soln) = 95.78H_2O(soln)$	$+0.34 \pm 0.02$
13. $2Cs(s) + Te(s) = Cs_2Te(s)$	-361.42 ± 3.20

Source: Cordfunke, E. H. P., and Ouweltjes, W. *J. Chem. Thermodyn.* 19, 377, 1987. With permission.

Example 2.10

The phenomena that can occur in the course of a nuclear reactor accident are complex because many chemical species and environments are involved. After the near-catastrophe in the nuclear power plant (TMI-2) in Harrisburg on Three Mile Island, USA, in 1979, investigators were very surprised to find that only a small fraction (10^{-6} to 10^{-7}) of the radioactive element Iodine-131 from the core material was released into the environment. After the catastrophe in the nuclear plant (RMBK-1000) in Chernobyl in Ukraine in 1986, this figure was 0.20. The minimal release of radioactive iodine in Harrisburg was due to the physical and chemical behavior of iodine and its compounds in the presence of water. However, before 1976, this was not known. After these accidents, much research has been carried out on the chemical changes that occur during accident conditions, where many gaseous species play a role in the transport of fission products.

Information about the chemical and physical forms of biologically important fission products is now becoming more readily available, with the emphasis on the volatile solids iodine, cesium, and tellurium. These form chemical compounds such as CsI and Cs_2Te, and are dissolved in the water within the enclosure of the reactor in a light-water nuclear plant such as in Harrisburg. Without going into details, the Chernobyl nuclear power plant used graphite as a moderator for neutrons, whereas the Harrisburg plant used light water as moderator. The circumstances of these accidents were therefore quite different, as were the chemical and physical processes that took place. Computer models are currently used to calculate the chemical equilibria in reactor accidents as a function of temperature and time to understand and predict these chemical phenomena. It is evident that the results of such analyses depend on the availability and accuracy of the thermodynamic data used.

Dicesium telluride, Cs_2Te, is an important compound in studies of the safety aspects of light-water nuclear reactors because it is formed during fission as a stable volatile compound. In Table 2.1, the reaction scheme for the enthalpy of formation $\Delta_f H^0$ of $Cs_2Te(s)$ is given. Often it is very difficult to carry out a measurement to estimate such data. By combining the first 12 reactions as in Table 2.1, it is possible to estimate the $\Delta_f H^0$ of Cs_2Te indirectly, according to Hess's law.

In these few examples, we have seen how very straightforward is the determination of ΔH for a reaction, either from formation data, or from related reactions. This is a skill that should become second nature to a student of thermodynamics, because such techniques also apply to free energy changes, ΔG_{298}^0 (Chapter 6), which can give invaluable information on chemical and other equilibria.

REFERENCES

1. Frisch, M. A., and J. L. Margrave. 1965. The heat of formation of nitric oxide(g). *J. Phys. Chem.* 69:3863–3866.
2. Cordfunke, E. H. P., and W. Ouweltjes. 1987. Standard enthalpies of formation of tellurium compounds II. Cs_2Te. *J. Chem. Thermodyn.* 19:377–379.
3. Gunn, S. R. 1966. The heat of formation of krypton difluoride. *J. Am. Chem. Soc.* 88:5924.
4. Vaughn, J. D., and E. L. Muetterties. 1960. Thermochemistry of sulfur tetrafluoride. *J. Phys. Chem.* 64:1787–1788.

PROBLEMS

2.1   The decomposition of krypton difluoride:

$$KrF_2(g) \rightarrow Kr(g) + F_2(g),$$

has been studied at 25°C by S. R. Gunn [3], who found that $\Delta U = -59.4$ kJ mol^{-1}.
(a)   Calculate ΔH for this reaction.
(b)   For sublimation of $KrF_2(s)$, $\Delta_{sub}H^0 = +41.4$ kJ mol^{-1}. Estimate $\Delta_f H_{298}$ [$KrF_2(s)$].

2.2   Gaseous normal propyl chloride may be hydrogenated to form propane and HCl gas; ΔH_{373}^0 is -65.7 kJ mol^{-1}. The related reaction with isopropyl chloride occurs with $\Delta H_{373}^0 = -58.37$ kJ mol^{-1}. What is ΔH_{373}^0 for the reaction:

$$iso\text{-}C_3H_7Cl \rightarrow n\text{-}C_3H_7Cl?$$

2.3   To determine the enthalpy of formation of CH_3Br more accurately, ΔH_{298}^0 for the reaction:

$$CH_3Br(g) + H_2(g) \rightarrow CH_4(g) + HBr(g)$$

has been determined as -74.49 kJ mol^{-1}. Use data from Appendix III to assist you in bringing this project to its conclusion.

2.4   Enthalpies of combustion for the different liquid isomers of hexane have been accurately determined as follows:

Isomer	$\Delta_c H^0$ (kJ mol^{-1})
n-Hexane	-4141.3
2-Methylpentane	-4135.9
3-Methylpentane	-4138.2
2,3-Dimethylbutane	-4133.2
2,2-Dimethylbutane	-4126.9

What are the enthalpy changes for isomerization, for the reaction

$$n\text{-hexane} \rightarrow \text{isomer}$$

in each case?

2.5   Dibrane (B_2H_6) has been considered as a high-energy rocket fuel. Assuming it burns to HBO_2(s) and H_2O(l), calculate the thermal energy available per gram of fuel-oxygen mixture. (This may be compared with a value of 8.24 kJ g^{-1} mixture for n-hexane.)

2.6   Acetic acid is currently manufactured by a high-temperature, high-pressure reaction:

$$CH_3OH(g) + CO(g) \rightarrow CH_3COOH(g).$$

Enthalpies of vaporization are 35.23 and 24.31 kJ mol^{-1} for methanol and acetic acid, respectively. Using data from Appendix III, estimate ΔH for this reaction. (*Hint:* Set up a reaction scheme such as in Example 2.10.)

2.7   Delrin is the DuPont trade name of a stabilized polyformaldehyde. The monomer is produced by partial oxidation of methanol to HCHO and H_2O at 600°C. What is ΔH^0_{298}?

The value of ΔH under plant conditions is -154.0 kJ mol^{-1}. Why is the agreement not exact?

2.8   The compound CHClF$_2$, an intermediate in the production of PTFE, is produced by the reaction:

$$CHCl_3(l) + 2HF(g) \rightarrow CHClF_2(g) + 2HCl(g).$$

Is this reaction endothermic?

2.9   Use Appendix III to assess the usefulness of propane and n-butane as fuels. Gram for gram, which provides more heat when burnt to CO_2 and liquid H_2O? (Molecular weights for H and C are 1.008 and 12.01 g mol^{-1}, respectively.)

2.10 Vaughn and Muetterties [4], have determined $\Delta H = -413.0$ kJ mol^{-1} SF$_4$ for the reaction:

$$SF_4(g) + 2H_2(g) \rightarrow S(s) + \tfrac{4}{n}(HF)_n(l).$$

This liquid HF product is partly polymerized, but it is known that the enthalpy of formation of $1/n(HF)_n$ when equally polymerized is $\Delta_f H = -282.8$ kJ/20 g HF (molecular weight of HF monomer is 20.0 g mol^{-1}).

Determine $\Delta_f H(SF_4(g))$.

3 Thermochemistry

In this chapter, we shall first look in more detail at the various experimental methods of measurement. These include determinations not only of heats of reaction, as ΔH or ΔU, but also of the *heat capacities* of individual materials. Armed with such information, we shall see that if the value of ΔH for a reaction is known at one temperature, it is possible to calculate it at another temperature. Finally, by using the concept of *bond energies*, we shall learn how to estimate values of $\Delta_f H$ for compounds that are rare or uncharacterized.

3.1 CALORIMETRY

This is one of the best established of the experimental sciences, and the ever-present challenge of more accurate determinations continues to draw the attention of research workers. Because reactions vary so much, with respect to reaction rates, physical states, corrosive qualities, and so on, there are very many types of calorimeters. However, they fall into a small number of broad classes. Because some calorimeters give ΔH directly, whereas others yield ΔU data, a short summary of types will follow.

3.1.1 BOMB CALORIMETERS

This type of apparatus is by far the most often used. The reactants, previously sealed in a strong container, are allowed to react under constant-volume conditions. Gases at high pressures are often used, hence the name. A line diagram of such an apparatus is shown in Figure 3.1.

The sequence of events during a typical experiment, in which an oxide is formed from a solid element and excess gas, is as follows. The element is carefully weighed and put in place, the igniting wire is positioned, and the bomb is sealed tightly. Oxygen is then pumped in at high pressure. When the temperature, T_1, of the system (that includes the water jacket) has been measured, a small electric current is passed to initiate the reaction. (The energy introduced through the fuse wire is compensated for later.) The temperature increase of the calorimeter, which is usually only a degree or two, is measured as accurately as possible (final temperature $= T_2$). Then, in a separate control experiment, the products plus calorimeter are raised from T_1 to T_2 by electrical heating, and the amount of electrical energy used is carefully measured. Thus, in the first case:

(A) System with reactants (T_1) → System with products (T_2),

and $q = w = 0$, so $\Delta U_A = q + w = 0$.

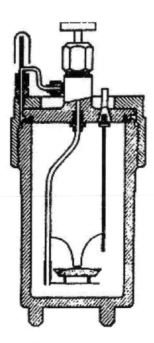

FIGURE 3.1 A typical calorimeter bomb for use with pure oxygen. The sample dish, igniting wire, and oxygen port are clearly visible. Not shown is the insulated water bath in which the bomb is immersed. There are facilities for stirring and accurately measuring the temperature of the water. (Reproduced from Jessup, R. S., *J. Res. Natl. Bur. Standards*, 21, 475, 1938. With permission.)

In the second case, electrical heating introduces Eit joules (E volts, i amps, flowing for t seconds), and so for:

(B) System with products (T_1) → System with products (T_2),

$$q = Eit, w = 0, \text{ and so } \Delta U_B = Eit.$$

Because U is a function of state, we can follow process (A) with the reverse of (B), giving:

System with products (T_1) → System with products (T_1),

for which,

$$\Delta U = \Delta U_A - \Delta U_B$$

$$= Eit \quad \text{J}.$$

That is to say that the electrical energy is equivalent to the energy "released" by the reaction when carried out at constant temperature T_1. By this means we may refer to

TABLE 3.1
A Selection of $\Delta_c H^0_{298}$ Values of Combustion

Material	State	Oxides	$\Delta_c H^0_{298}$ of Combustion (kJ mol^{-1})
Al	s	Al_2O_3	−1675.7
Ca	s	CaO	−634.9
Cr	s	Cr_2O_3	−1140.6
Mo	s	MoO_3	−744.6
Re	s	Re_2O_7	−1248.5
Si	s	SiO_2	−910.9
C_6H_6	l	CO_2, H_2O	−3273
C_2H_6	g	CO_2, H_2O	−1560.5
CH_4	g	CO_2, H_2O	−890.6
$(C_2H_5)_2O$	l	CO_2, H_2O	−2727

ΔU at *constant* temperature, usually 298 K, even though instantaneous temperatures of the reacting compounds would be very much higher for short periods. Numerous corrections must be applied before an accurate value of ΔU may be calculated. This is usually then converted to a value for ΔH^0_{298}, the standard enthalpy change for the reaction, using the method described in Section 2.2.1. For the more straightforward reactions, a precision of 1 in 10,000 is not only possible, but necessary. In many cases, however, combustion is incomplete, or slow, or otherwise difficult, and precision is substantially impaired.

Table 3.1 shows a selection of ΔH^0_{298} values for the combustion of a number of elements and organic compounds. (In the case of the elements, the values represent enthalpies of formation of the oxides.)

Example 3.1

Cobble et al. [1] performed the reaction:

$$2Tc(s) + \tfrac{7}{2}O_2(g) + H_2O(l) + aq \rightarrow 2HTcO_4(\text{dil. solution})$$

in a bomb calorimeter.

In a typical experiment, 56.39 mg of technetium caused a temperature rise of 0.0572°C on combustion. Electrical calibration of the calorimeter shows its "water equivalent" to be 5847 J K^{-1}, so this experiment gives a value of:

$$\Delta U = (-0.0572)(5847) = -334.45 \text{ J}/56.39 \text{ mg Tc.}$$

If the atomic weight of Tc is 98.80 g mol^{-1}, then:

$$\Delta U = (-334.45)2(98.80)/(56.39) \times 10^{-3} = -1172 \text{ kJ}/2 \text{ mol Tc.}$$

This leads to a value of $\Delta H = -1180$ kJ/2 mol Tc (which should be checked by the student). Moreover, the heat of (dilute) solution of Tc_2O_7 to form dilute pertechnic acid is known to be $\Delta H = -51.0$ kJ mol^{-1} Tc_2O_7. Using this datum, it is then possible to calculate the standard heat of formation of Tc_2O_7(s), which turns out to be $\Delta_f H^0(Tc_2O_7(s)) = -1129$ kJ mol^{-1}. (This, too, should be checked.)

Fluorine bomb calorimetry is a development from the early 1960s. Before that time, reliable enthalpy data concerning fluorides were very scarce, principally because fluorine gas is so very reactive. Fluorine bomb calorimetry was extended to high-pressure (up to 15 atm of fluorine) metal combustion bombs by Greenberg and coworkers [2] at the Argonne National Laboratory in the United States in 1961. The technique has been developed over the past 70 years, and is now comparable in precision and accuracy to the other types of calorimetry. Enthalpies of formation have been determined from direct fluorination experiments.

Example 3.2

In a study by Cordfunke and Ouweltjes [3], based on solution calorimetric measurements, $\Delta_f H^0(UF_4(s))$ was more negative than suggested in previous literature. Therefore, it was decided to obtain independent measurements of the enthalpy of formation of UF_4 using fluorine calorimetry [4].

The reaction of UF_4 with fluorine:

$$UF_4(s) + F_2(g) \rightarrow UF_6(s) \qquad \text{I}$$

was chosen because this reaction can be forced to go to completion by using an auxiliary combustion aid.

$$\text{Measured: } \Delta_c U^0_{298} = -278.5 \text{ kJ mol}^{-1}.$$

From this value, $\Delta_c H^0 = -281.0$ kJ mol^{-1} can be derived. (This should be checked.)

The auxiliary reaction:

$$U(s) + 3F_2(g) \rightarrow UF_6(s) \qquad \text{II}$$

required to derive the enthalpy of formation of UF_4, was studied at Argonne National Laboratory [5]. From this study and the previous one [4], it was concluded that $\Delta_f H^0(UF_6(c)) = -2197.2$ kJ mol^{-1}.

The equation for the formation of uranium tetrafluoride:

$$U(s) + 2F_2(g) \rightarrow UF_4(s) \qquad \text{III}$$

is the difference between II and I, so:

$$\Delta_f H^0_{298}(UF_4(s)) = -2197.2 + 281.0 = -1916.2 \text{ kJ mol}^{-1}.$$

3.1.2 Differential Scanning Calorimetry

Thermal analysis is the measurement of a given characteristic of a substance as a function of time. One of the most popular techniques today in thermal analysis is differential scanning calorimetry (DSC), in which the behavior of a sample is determined by comparing it with an inert reference, which consequently shows no thermal activity in the temperature range of interest. It is obvious that, in this way, changes in the temperature of the sample can be detected much more accurately than when only the absolute temperature of the sample is measured.

In so-called power-compensation DSC, sample and reference are completely isolated from each other (Figure 3.2). Both the sample and reference crucible have their own heating element and temperature-sensing element. With the aid of a temperature programmer, both sample and reference are heated and always have the same temperature. As soon as changes in the sample occur, extra (or less) heat will be needed to maintain the set heating rate. With the aid of specialized electronic circuitry, extra (or less) power is now sent to the sample holder to keep the temperature difference zero. In this manner, power and consequently heat flow and enthalpy changes are measured.

DSC instruments are sensitive pieces of modern equipment, having the capability to measure heat flows of the order of microwatts. This feature makes the applicability of the technique almost unlimited: every physical change or chemical reaction takes place with a change of enthalpy and consequently absorption or release of heat.

A characteristic DSC curve is shown in Figure 3.3. The temperature is always plotted on the x-axis. In this figure heat flow is represented by $\dot{q}$, defined as $\Delta q/\Delta t$, the amount of heat flowing per unit of time. DSC curves may be used solely for "fingerprint" comparison with sets of reference curves. It is usually possible, however, to extract a great deal more information from the curves, such as temperatures and enthalpy changes for the thermal events occurring. As an example, the DSC curve for $CaSO_4 \cdot 2H_2O$ is shown in Figure 3.3. The area under the endotherm is related to the value of the enthalpy change, ΔH, for the thermal event.

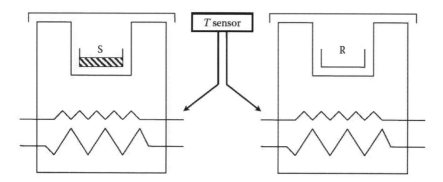

FIGURE 3.2 Schematic drawing of a power-compensation DSC cell. S, sample; R, reference; T, temperature.

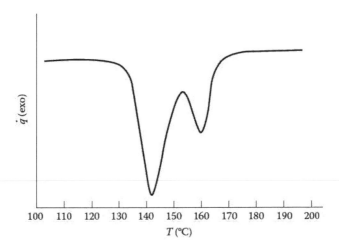

FIGURE 3.3   DSC curve for $CaSO_4 \cdot 2H_2O$.

The $CaSO_4$–H_2O equilibrium is as follows:

$$CaSO_4 \cdot 2H_2O \rightleftharpoons CaSO_4 \cdot \tfrac{1}{2}H_2O + 1\tfrac{1}{2}H_2O$$

$$CaSO_4 \cdot \tfrac{1}{2}H_2O \rightleftharpoons CaSO_4 + \tfrac{1}{2}H_2O.$$

The positions of these equilibria are determined by the partial pressure of water vapor. When a $CaSO_4 \cdot 2H_2O$ sample is heated in an open crucible, both reactions will proceed rapidly when the partial pressure of water exceeds 1 bar and the DSC curve will show one peak. When, however, heating is effected in an almost entirely closed system, the vapor pressure may become so high that the two reactions occur one after the other and two peaks are shown. Explain why the ratio of the peak areas in Figure 3.3 is about 3:1.

3.2   CONCEPTS OF HEAT CAPACITY

In many respects, water is an exceptional material. The specific heat of water is one exceptional property for which we all have cause to be thankful. The high specific heat of water is responsible for, among other things, the comfort of a hot water bottle on a winter's night, and the equable climate of oceanic regions. Questions of specific heat concern us frequently. The property we shall use, rather than the specific heat, is the *molar heat capacity*, which is the energy required to raise the temperature of 1 mol of a substance by 1 K. Defined in terms of a differential, the heat capacity is:

$$C = \frac{\mathrm{d}q}{\mathrm{d}T}, \quad \mathrm{J \ mol^{-1} \ K^{-1}}.$$

As it stands, this definition is not complete, because $\mathrm{d}q$ needs to be specified. This $\mathrm{d}q$ depends on the actual process followed. If the volume is constant:

$$C_V = \frac{\mathrm{d}q_V}{\mathrm{d}T} = \frac{\mathrm{d}U}{\mathrm{d}T}.$$

At constant pressure:

$$C_P = \frac{\mathrm{d}q_P}{\mathrm{d}T} = \frac{\mathrm{d}H}{\mathrm{d}T}.$$

We recall that U and H are closely linked, and so it follows that C_V and C_P must also be linked. That is:

$$C_P = \frac{\mathrm{d}H}{\mathrm{d}T} = \frac{\mathrm{d}(U + PV)}{\mathrm{d}T} = \frac{\mathrm{d}U}{\mathrm{d}T} + \frac{\mathrm{d}(PV)}{\mathrm{d}T}.$$

For 1 mol gas that behaves ideally, $PV = RT$, and $\mathrm{d}(PV) = R\mathrm{d}T$. Thus:

$$C_P = C_V + R \quad \text{(ideal gas).} \tag{3.1}$$

The case of solids and liquids is not as simple. The equation of state is more complex, due to what is known as *internal pressure*, which arises from close-range interatomic forces. These are neglected in the case of ideal gases.

For gases, the heat capacity increases both with the mass of each atom in the molecule, and with the degree of molecular complexity. Values of entropy, to be discussed in Chapter 5, are closely linked to heat capacity, and show similar qualitative variations. Let us now suppose that we are asked to calculate the thermal energy required to heat a sample of methyl chloride (at constant pressure) over a range of temperatures, say from T_1 to T_2. From the definition of C_P.

$$\mathrm{d}q_P = C_P\mathrm{d}T,$$

and so

$$q_P = \int_{T_1}^{T_2} C_P\mathrm{d}T. \tag{3.2}$$

There are three possible courses of action:

(a) Temperature range is small.
In general, C_P varies with temperature, but for a small change, we can take an average value of C_P, and then:

$$q_P = C_P \int_{T_1}^{T_2} dT = C_P(T_2 - T_1) \text{ J mol}^{-1}.$$

(b) Values of C_P are known in tabular form.
The so-called JANAF tables [6] are an excellent compilation of thermodynamic data in tabular form. Table 3.2 is an extract of the heat capacity data for methyl chloride over a range of temperatures. By drawing a simple graph of C_P against T, and measuring the area, $q_P = \int C_P dT$ may easily be determined. From Figure 3.4 it is found that 41.76 kJ are needed to heat 1 mol CH_3Cl from 350 to 1000 K.

(c) C_P is known as a power series.
As can be seen from Figure 3.4, C_P varies in a nonlinear fashion with temperature. The number of terms commonly used to describe C_P varies from two to four. High-temperature reactions are becoming increasingly important, and as a result wider variations in C_P are encountered, so we shall use expressions of the form:

$$C_P = a + bT + cT^2 + dT^3. \tag{3.3}$$

The terms a, b, c, and d are constant for a particular material, and T is in K. For methyl chloride:

$$C_P = 12.76 + 10.86 \times 10^{-2}T - 5.205 \times 10^{-5}T^2 + 9.623 \times 10^{-9}T^3,$$

TABLE 3.2

Heat Capacity Data for CH_3Cl, from 300 to 1000 K

T (K)	C_P (J mol^{-1} K^{-1})
300	40.82
400	48.10
500	55.09
600	61.25
700	66.60
800	71.26
900	75.33
1000	78.90

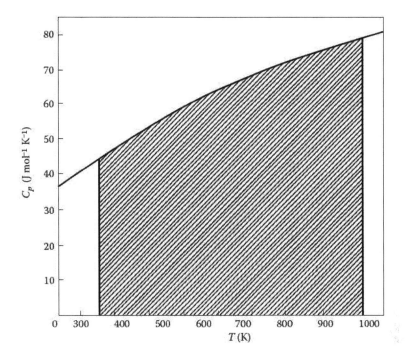

FIGURE 3.4 Determination of thermal energy needed to raise 1 mol of CH_3Cl from 350 to 1000 K.

and this is valid over the range 273–1500 K. In this case, $a = 12.76$, and so on. Table 3.3 gives data for a selection of materials, in terms of Equation 3.3.

To calculate q_P, we must use Equation 3.2:

$$q_P = \int_{T_1}^{T_2} C_P dT,$$

Using Equation 3.3, this becomes:

$$q_P = \int_{T_1}^{T_2} \left(a + bT + cT^2 + dT^3\right) dT$$

$$= \left[aT + \frac{bT^2}{2} + \frac{cT^3}{3} + \frac{dT^4}{4}\right]_{T_1}^{T_2}.$$

TABLE 3.3

Molar Heat Capacities for a Selection of Gases According to the Equation
$C_p = a + bT + cT^2 + dT^3$

Compound		a	10^2b	10^5c	10^9d	Valid (K)
Acetylene	C_2H_2	20.05	9.951	−6.819	16.92	273–2000
Ammonia	NH_3	27.55	2.563	0.9900	−6.686	273–1500
Benzene	C_6H_6	−36.19	48.44	−31.55	77.57	273–1500
Carbon monoxide	CO	27.11	0.655	−0.1000	–	273–3800
Carbon dioxide	CO_2	22.24	5.979	−3.498	7.464	273–1800
Ethane	C_2H_6	6.895	17.25	6.402	7.280	273–1500
Ethanol	C_2H_5OH	19.9	20.95	−10.372	20.04	273–1500
Ethylene	C_2H_4	3.95	15.63	−8.339	17.66	273–1500
Hydrogen	H_2	29.09	−0.1916	0.4000	−0.870	273–1800
Hydrogen fluoride	HF	30.13	−0.494	0.6594	−1.573	273–2000
Hydrogen chloride	HCl	30.31	−0.761	1.326	−4.335	273–1500
Methane	CH_4	19.87	5.021	1.268	−11.00	273–1500
Methyl chloride	CH_3Cl	12.76	10.86	−5.205	9.623	273–1500
Nitrogen	N_2	27.32	0.6226	−0.0950	–	273–3800
Oxygen	O_2	25.46	1.519	−0.7150	1.311	273–1800
Water	H_2O	29.16	1.449	−0.2022	–	273–1800

Source: Kobe, K.A., *Petroleum Refiner*, November 1954. With permission. Units are in J mol^{-1} K^{-1}.

Using methyl chloride as an example, with data from Table 3.3, we have:

$$q_P = \left[12.76T + \frac{10.86 \times 10^{-2}T^2}{2} - \frac{5.205 \times 10^{-5}T^3}{3} + \frac{9.623 \times 10^{-9}T^4}{4} \right]_{350}^{1000}$$

$$= 8297 + 47{,}650 - 16{,}602 + 2368$$

$$= 41.71 \text{ kJ mol}^{-1}.$$

This value agrees well with the previous graphical value of 41.76 kJ mol^{-1}. However, at higher temperatures Equation 3.3 no longer holds. The following equation corresponds better with reality:

$$C_P = a + bT + cT^2 + \frac{d}{T^2} \quad (\text{in J mol}^{-1} \text{ K}^{-1}).$$

In this equation, the constants a, b, c, and d are only valid in certain temperature ranges. These values are listed in the Scientific Group Thermodata Europe database, from which the data of Appendix III are taken. For reasons of simplicity, we will make use of Equation 3.3.

3.2.1   COMBUSTION AND FLAME TEMPERATURES

Flames are exceedingly complex. As a result, it would seem that to attempt to calculate the temperature of, say, the combustion gases in a turbojet engine would be quite useless. It is not impossible, however, as we shall see. Moreover, by making one or two intelligent simplifications, we shall be able to determine a *theoretical maximum flame temperature*, which, although it will need to be carefully qualified, will be a starting point for further analysis. But first, picture a high-temperature flame. The premixed gases are completely reacted in a few milliseconds, rapidly reaching a high temperature.

Many chemical intermediates and radicals are present, and radiation of energy occurs immediately after the reaction is initiated. Some actual flame temperatures are given in Table 3.4, and serve to show the range of nearly 3000 degrees that is commonly met. To add to the apparent complications, we must remind ourselves that ΔH for a reaction changes slightly with temperature, and so also does C_P for the gaseous products, which absorb much of the available thermal energy. However, the first and principal simplification is to apply the first law. If we assume an adiabatic process (no heat lost from the flame itself, either by conduction or radiation), then we can use the fact that H is a function of state. The overall process is:

Fuel + oxidant (298K) $\rightarrow$ Products (several thousand degrees),

but it may be performed in two (imaginary) steps.

I   Fuel + oxidant (298 K) $\rightarrow$ Products (298 K) ($\Delta H < 0$)
II Products (298 K) $\rightarrow$ Products (several thousand degrees) ($\Delta H > 0$).

That is, the thermal energy available from the exothermic reaction is then used to heat the products from 298 K to their final temperature. If we use Equation 3.2, which is:

$$q_P = \int_{T_1}^{T_2} C_P dT,$$

TABLE 3.4
Maximum Flame Temperatures Attainable by Using Various Fuel Gases

Fuel/Oxidant Mixture	Maximum Temperature (K)
Propane-air	2200
Hydrogen-air	2320
Acetylene-oxygen	3430
Cyanogen-oxygen	4910

we have to solve for T_2, knowing q_p (= $|\Delta H|$ of the reaction), T_1 (= 298 K), and C_p (of the products). One last point should be made before we make a sample calculation. The "products" must be carefully specified. Thus, of the common flame products, CO_2 dissociates extensively above 1000 K into CO and O_2, and water breaks up to give OH radicals and O and H atoms above about 3500 K.

Example 3.3

Over many years, the simple "flame test," whereby atoms of, for instance, sodium are excited in a flame to give a characteristic yellow color, has been developed as a sophisticated and sensitive instrumental technique (flame emission spectro-photometry). Sensitivity depends on dissociation of the injected materials into free atoms such that the characteristic atomic emissions can be given. This, in turn, demands high flame temperatures. The combination of acetylene fuel with nitrous oxide as oxidant has proved highly successful for this purpose. What temperature is possible in theory?

The best mixture would correspond to the equation:

$$C_2H_2 + 3N_2O \rightarrow 2CO + H_2O + 3N_2.$$

(If more N_2O were used, would CO_2 be formed?)

Using enthalpies of formation from Appendix III we have (using Equation 2.10):

$$\Delta H^0 = 2(-110.5)+(-241.8)-(227.4)-3(81.60)$$

$$= -935.0 \text{ kJ.}$$

Notice that the enthalpies of formation of both C_2H_2 and N_2O are positive, and energy is released upon their breakdown. This energy is now available for heating the gaseous products. (We shall assume that water is undissociated.)

Now, $q_p = \int_{T_1}^{T_2} C_p dT$ becomes in this case:

$$+935,000 = \int_{298}^{T} C_p(2CO+H_2O+3N_2)dT.$$

By determining C_p for each product gas, and adding, C_p for the product is found to be:

$$C_{p,\text{Products}} = 2C_p(CO)+C_p(H_2O)+3C_p(N_2)$$

$$= 165.3+4.6268\times10^{-2}T-0.6872\times10^{-5}T^2.$$

Therefore,

$$93,500 = \int_{298}^{T} \left[165.3 + 4.6268 \times 10^{-2}T - 0.6872 \times 10^{-5}T^2 \right] dT.$$

On integrating this expression, and inserting the limits, we have the cubic equation:

$$22.907 \times 10^{-7}T^3 - 23.134 \times 10^{-3}T^2 - 165.3T + (986,253) = 0$$

from which we find that $T = 4426$ K. The result is encouragingly high and the nitrous oxide–acetylene flame is widely used. Temperatures of 3230 K are achieved in practice, and this shows that significant errors arise from neglecting water decomposition and radiation losses.

3.2.2 VARIATION OF REACTION ENTHALPIES WITH TEMPERATURE

Suppose that a new vapor-phase hydrogenation process is to be launched on a pilot-plant scale. The reaction equation is known, and so ΔH^0_{298} is easily calculated from formation data. However, the reaction proceeds at a convenient rate at, for example, 600 K. Will ΔH^0 change? By how much? In which direction?

Heat capacities are clearly involved in the heating of reactants and cooling of products between 600 and 298 K. The simplest analysis is as follows. The change of ΔH with temperature is:

$$\frac{d(\Delta H)}{dT} = \frac{d\left(H_{products} - H_{reactants}\right)}{dT}$$

$$= \left(\frac{dH}{dT}\right)_{products} - \left(\frac{dH}{dT}\right)_{reactants}$$

$$= C_P(products) - C_P(reactants).$$

This simple result is often written

$$\frac{d(\Delta H)}{dT} = \Delta C_P. \tag{3.4}$$

In an exactly analogous way, it can be shown that:

$$\frac{d(\Delta U)}{dT} = \Delta C_V \tag{3.5}$$

These equations are often known as Kirchhoff's equations, after their first protagonist. Equation 3.4 is more widely used than Equation 3.5, and our attention will be confined to it from now on.

Equation 3.4 may be integrated:

$$\int_{T_1}^{T_2} d(\Delta H) = \Delta C_p dT,$$

(3.6)

$$\Delta H_2 - \Delta H_1 = \int_{T_1}^{T_2} \Delta C_p dT.$$

Normally, ΔC_P will take the form of a combination of C_P expressions in T, which must then be integrated. For small temperature ranges, mean values for C_P may be taken, and then:

$$\Delta H_2 = \Delta H_1 + \Delta C_P(T_2 - T_1).$$

(3.7)

Example 3.4

Using a flow calorimeter, Lacher et al. [7] have studied the hydrogenation of methyl chloride:

$$CH_3Cl + H_2 \rightarrow CH_4 + HCl.$$

They found that, at 248°C (521 K), $\Delta H_{521}^0 = -82.32$ kJ mol^{-1}. Before this result can be incorporated into tables of standard data, adjustment to the standard temperature (298 K) must be made. For this calculation we will use values of C_p at 400 K, of:

$$C_P(CH_4(g)) = 40.66 \text{ J mol}^{-1} \text{ K}^{-1},$$

$$C_P(HCl(g)) = 29.15 \text{ J mol}^{-1} \text{ K}^{-1},$$

$$C_P(CH_3Cl(g)) = 48.21 \text{ J mol}^{-1} \text{ K}^{-1},$$

$$C_P(H_2(g)) = 29.16 \text{ J mol}^{-1} \text{ K}^{-1}.$$

Use Equation 3.7, with "1" denoting the experiment at 521 K, and "2" denoting the process at 298 K. Then:

$$\Delta C_P = C_P(\text{products}) - C_P(\text{reactants})$$

$$= 40.66 + 29.15 - 48.21 - 29.16$$

$$= -7.56 \text{ J mol}^{-1} \text{ K}^{-1},$$

and so:

$$\Delta H_{298}^0 = \Delta H_{521}^0 + (-7.56)(298 - 521)/1000$$

$$= -82.32 + 1.686$$

$$= -80.63 \text{ kJ mol}^{-1}$$

Thus, we see that the reaction would be nearly 2% less exothermic if it were to be carried out at 298 K. As a general guide, although the heat capacities tend to be self-cancelling, there are significant changes in ΔH that must be accounted for, especially when considering reactions in solution.

3.3   BOND ENERGIES

A chemical reaction often involves the breaking of some bonds, and the making of others. It seems feasible, therefore, that there should be a close link between the stability of a bond and the enthalpy change involved in breaking it. We can represent the breaking of the X—Y bond in 1 mol of the molecule XY (not necessarily diatomic) by the equation:

$$XY(g) \rightarrow X(g) + Y(g) \quad \text{enthalpy change} = \Delta H_{298}^0 .$$

If the fragments and the compound are gases in their standard states, at 1 bar and 25°C, then we can define the *standard bond dissociation energy* (BDE), DH^0 for the bond X—Y as:

$$DH^0(X - Y) = \Delta H_{298}^0 .$$

For a particular bond, the value of DH^0 will depend on the local environment, the presence or absence of nearby electrophilic groups, and so on. For example, the two O—H bonds of water can be broken in succession:

$$HO - H(g) \rightarrow HO(g) + H(g), \quad \Delta H = +494 \text{ kJ mol}^{-1}$$

$$O - H(g) \rightarrow O(g) + H(g), \quad \quad \Delta H = +431 \text{ kJ mol}^{-1}.$$

The standard BDE in each case is therefore $DH^0(HO—H) = 494$ kJ, and $DH^0(O—H) = 431$ kJ. Methane provides a second, more extreme example. The successive removal of hydrogen atoms may be summarized:

$$DH^0(CH_3 - H) = 435 \text{ kJ},$$

$$DH^0(CH_2 - H) = 444 \text{ kJ},$$

$$DH^0(CH - H) = 444 \text{ kJ},$$

$$DH^0(C - H) = 339 \text{ kJ}.$$

It is perhaps surprising not that the values vary, but that they are as close as they are. Values of DH^0 for a small number of specific bonds have been determined with difficulty using such techniques as mass spectrometry and ultraviolet and visible spectroscopy. On the whole, however, accurate values are rather few and far between, and as there are far more bonds than there are chemical compounds, the prospect for complete tables of DH^0 is slight indeed.

3.3.1 AVERAGE BOND DISSOCIATION ENERGIES

An alternative approach is to consider the complete atomization of a compound, such as methane, into its constituent atoms in the gas phase:

$$CH_4(g) \rightarrow C(g) + 4H(g).$$

This reaction involves the breaking of all four C—H bonds, and so the enthalpy change will be a sum of all four standard bond dissociation energies. However, we can write, from Equation 2.10:

$$\Delta H^0 = \Delta_f H^0(C(g)) + 4\Delta_f H^0(H(g)) - \Delta_f H^0(CH_4(g)). \tag{3.8}$$

Enthalpies of atomization, or formation of atoms, are much easier to determine than specific bond energies, and are, on the whole, more accurate.

Values for several important elements are included in Appendix III. On substituting values into Equation 3.8:

$$\Delta H^0 = 716.7 + 4(218.0) - (-74.60)$$

$$= 1663 \text{ kJ}.$$

The *average* BDE, DH_{av}, can now be calculated. The above value represents the breaking of four C—H bonds:

$$DH_{av}(C—H) = \frac{\Delta H^0}{4} = \frac{1663}{4} = 415.8 \text{ kJ mol}^{-1}.$$

Let us now assume that for ethane, the six C—H bonds have an identical average DBE. Atomization of ethane involves rupture of six C—H bonds, and one C—C bond. Now, ΔH^0 for ethane atomization, which can be calculated as before, is:

$$\Delta H^0 = 2\Delta_f H^0(C(g)) + 6\Delta_f H^0(H(g)) - \Delta_f H^0(C_2H_6(g)),$$

which also

$$= 6DH_{av}(C—H) + DH_{av}(C—C),$$

TABLE 3.5
Values of DH_{av} (C—H) and DH_{av} (C—C) Calculated from Normal Alkanes (in kJ mol^{-1})

n-Alkane	DH_{av} (C—H)	DH_{av} (C—C)
CH_4	415.5	–
CH_4, C_2H_6	415.5	329.7
C_2H_6, C_3H_8	413.0	343.9
C_3H_8, C_4H_{10}	411.3	350.2
C_4H_{10}, C_5H_{12}	413.0	345.2
C_5H_{12}, C_6H_{14}	412.5	346.4

and so a value for the C—C bond may be calculated. In this way, successive pairs of hydrocarbons yield a series of values of DH_{av} for the C—H and C—C bonds. These are summarized in Table 3.5. It is clear that such bonds are of equal strength in whatever molecule they occur.

Although "irregular" environments will cause serious discrepancies, it is nonetheless possible to use such average values to predict properties of compounds of a given structure. Such average bond energies have been determined for many other bonds, and they provide a rough and ready guide to heats of formation, and of combustion, if these are otherwise unknown. Table 3.6 summarizes such values.

TABLE 3.6
Average Bond Dissociation Energies for a Selection of Single and Multiple Bonds (in kJ mol^{-1})

	I	Br	Cl	F	O	C	H
H	297	368	431	569	460	414	435
C	226	289	343	444	339	343	
O	–	–	205	184	146		
F	243	251	255	159			
Cl	209	218	243				
Br	180	192					
I	151						

C = C	611 kJ mol^{-1}
C ≡ C	833 kJ mol^{-1}
C = O	741 kJ mol^{-1}
C ≡ N	879 kJ mol^{-1}

Example 3.5

Ethyl hydroperoxide $C_2H_5OOH(g)$ is rumored to have a value of $\Delta_f H^0 = -167.4$ kJ mol^{-1}. Is this feasible?

First, let us write down the atomization process, as:

$$C_2H_5OOH(g) \rightarrow 2C(g) + 6H(g) + 2O(g).$$

This involves breaking of five C—H, one C—C, one C—O, one O—O, and one O—H bond, and so, from Table 3.6:

$$\Delta H^0 = 5(414) + 343 + 339 + 146 + 460 = 3358 \text{ kJ mol}^{-1}.$$

In addition, we can use enthalpies of formation, and:

$$\Delta H^0 = 2\Delta_f H^0(C(g)) + 6\Delta_f H^0(H(g)) + 2\Delta_f H^0(O(g)) - \Delta_f H^0(C_2H_5OOH(g))$$

$$= 2(716.7) + 6(218.0) + 2(249.2) - \Delta_f H^0(C_2H_5OOH(g)).$$

By eliminating ΔH^0, we find the value:

$$\Delta_f H^0(C_2H_5OOH) = 3239 - 3358 = 118 \text{ kJ mol}^{-1},$$

which is reasonable, considering that it is a small difference between two large figures.

Example 3.6

Vaughn and Muetterties [8] found that $\Delta H = -718.4$ kJ mol^{-1} for the formation reaction:

$$S(s) + 2F_2(g) \rightarrow SF_4(g).$$

What is the average BDE of the S—F bond in SF_4?

Complete bond dissociation implies the reaction:

$$SF_4(g) \rightarrow S(g) + 4F(g),$$

for which,

$$4DH_{av}(S—F) = \Delta H^0$$

$$= \Delta_f H^0(S(g)) + 4\Delta_f H^0(F(g)) - \Delta_f H^0(SF_4(g))$$

$$= 277.2 + 4(79.4) - (-718.4)$$

$$= 1313$$

$$DH_{av}(S—F) = \frac{1314}{4} = 328.3 \text{ kJ mol}^{-1}.$$

REFERENCES

1. Cobble, J. W., W. T. Smith Jr., and G. E. Boyd. 1953. Thermodynamic properties of technetium and rhenium compounds. *J Am Chem Soc* 75:5777–5782.
2. Greenberg, E., J. L. Settle, H. M. Feder, and W. N. Hubbard. 1961. Fluorine bomb calorimetry. I. The heat of formation of zirconium tetrafluoride. *J Phys Chem* 65:1168–1172.
3. Cordfunke, E. H. P., and W. Ouweltjes. 1981. Standard enthalpies of formation of uranium compounds VII. UF_3 and UF_4 (by solution calorimetry). *J Chem Thermodyn* 13:193–197.
4. Wijbenga, G. 1981. Thermochemical investigations on intermetallic UMe3 compounds. Thesis, University of Amsterdam, The Netherlands.
5. Johnson, G. K. 1979. The enthalpy of formation of hexafluoride. *J Chem Thermodyn* 11:483–490.
6. *JANAF Thermochemical Tables*, Publication PB 168–370. Springfield, VA: Clearinghouse for Federal Scientific and Technical Information.
7. Lacher, J. R., E. Emery, E. Bohmfalk, and J. D. Park, 1956. Reaction heats of organic compounds. *J Phys Chem* 60:492–495.
8. Vaughn, J. D., and E. L. Muetterties. 1960. Thermochemistry of sulfur tetrafluoride. *J Phys Chem* 64:1787–1788.

PROBLEMS

3.1 A sample of 39.44 mg liquid phenyl acetylene, C_6H_5CCH (mol. wt. = 102.14 g mol^{-1}) is burned in pure oxygen in a bomb calorimeter at 25°C. Thermal energy is released, equivalent to 1656 J of electrical energy. What is, per mole, q_v, w, ΔU? Assuming that the products are $CO_2(g)$ and $H_2O(l)$, calculate ΔH.

3.2 In a fluorine bomb calorimeter at 298 K, the combustion of $UF_3(s)$ to give $UF_6(s)$ resulted in $\Delta_c U^0 = -2342.5$ J g^{-1} UF_3. Estimate $\Delta_c H^0$ per mole UF_3. (The molar mass of UF_3 is 295.0 g mol^{-1}.)

   If given $\Delta_f H^0_{298}(UF_6) = -2197.2$ kJ mol^{-1} [3], calculate $\Delta_f H^0_{298}(UF_3)$.

3.3 A flow calorimeter is operated at 1 atm and 248°C. In the formation of HCl gas from H_2 and Cl_2, a typical run showed that $(3.913)10^{-4}$ mol HCl min^{-1} liberated 36.56 J min^{-1} of energy.

   What is the $\Delta_f H^0$(HCl) at 248°C? If average values of C_P are:

$$HCl(g): 29.1 \text{ J mol}^{-1} \text{ K}^{-1},$$

$$H_2(g): 28.8 \text{ J mol}^{-1} \text{ K}^{-1},$$

$$Cl_2(g): 33.9 \text{ J mol}^{-1} \text{ K}^{-1},$$

   determine the $\Delta_f H^0_{298}(HCl)$.

3.4 The Sachse process is used for the production of acetylene from natural gas:

$$2CH_4 \xrightleftharpoons{1500°C} C_2H_2 + 3H_2$$

From Appendix III, ΔH^0_{298} is found to be +376.6 kJ mol^{-1} C_2H_2. Using the following average values of C_P, estimate ΔH^0 at the operating temperature:

$$C_p(CH_4) = 61.6 \text{ J mol}^{-1} \text{ K}^{-1},$$

$$C_p(C_2H_2) = 62.7 \text{ J mol}^{-1} \text{ K}^{-1},$$

$$C_p(H_2) = 31.2 \text{ J mol}^{-1} \text{ K}^{-1}.$$

3.5   The use of average heat capacities in the previous problem leads to poor accuracy, especially over the very large temperature range involved. Using $C_p(T)$ functions from Table 3.3, make a better estimate of the actual ΔH^0 value at 1500°C.

3.6   Dungeness B nuclear station is centered around the first large "advanced gas-cooled reactor" (AGR). The primary coolant for the uranium oxide fuel elements is high-pressure CO_2 gas. It enters the core at 561 K and leaves at 918 K. It is obviously essential to know the amount of CO_2 needed to remove 1200 megawatts from the core. Use C_p data to find
(a) $(H_{918}—H_{561})$ in kJ mol^{-1}, algebraically or graphically
(b) How many kg of CO_2 must pass per second?

3.7   Mercury boils at 629.7 K under atmospheric pressure, and $\Delta_{vap}H = 59.27$ kJ mol^{-1} at that temperature. A mercury distillation plant runs at 673.0 K; what is $\Delta_{vap}H$ at that temperature? Average values of C_p for liquid and vapor are 27.4 and 20.6 J mol^{-1} K^{-1}, respectively.

3.8   Ammonia fuel with pure fluorine oxidant gives a very high-temperature flame:

$$NH_3 + \tfrac{3}{2}F_2 \rightarrow \tfrac{1}{2}N_2 + 3HF.$$

Assuming that the molar heat capacities of the products may be averaged over the temperature range to:

$$C_p(N_2) = 32.55 \text{ J mol}^{-1} \text{ K}^{-1},$$

$$C_p(HF) = 31.60 \text{ J mol}^{-1} \text{ K}^{-1},$$

calculate the theoretical flame temperature starting at 25°C. Neglect the effects of radiation losses and product dissociation.

3.9   For ethylene burning with pure oxygen to give CO and H_2O, the theoretical flame temperature approaches 5000 K. Make an accurate calculation of this for reactants at 25°C, using the C_p equations from Table 3.3. (This will involve solution of a cubic equation, which can be done with a scientific pocket calculator or by means of a computer program.)

3.10 Compare the value of -134.5 kJ mol^{-1} for the standard enthalpy change of formation of gaseous 2-methylpropane (C_4H_{10}) with that calculated from bond energy data (Table 3.6 and Appendix III).

3.11 The standard enthalpy change for formation of $KrF_2(g)$ is $+ 60.20$ kJ mol^{-1}. Using atomization data from Appendix III, determine the bond dissociation energy for the Kr—F bond.

4 Spontaneous Changes

So far, we have learned to follow, by means of the first law, the transfer of energy and its conversion from one form to another. A question that is often asked is, "Will this particular process tend to occur or not?" The first law provides no answer to this type of question, and we must look further for a guide to reactivity. By considering various spontaneous processes, we shall find that there are two driving forces in nature. The drive toward minimization of energy is one such directing influence, but there is also a tendency for materials to become more physically disorganized. The concept of entropy will be treated in more detail in Chapter 5, but is introduced here as a measure of the randomness of materials; the tendency for entropy to increase is nature's second driving force.

4.1 EVERYDAY PROCESSES

Our daily lives consist of a multitude of more or less familiar occurrences, most of which are so common that we do not stop to consider the direction of these changes. Water runs downhill; a plate, dropped onto the floor, breaks and the pieces scatter; petroleum vapor and oxygen, when reminded by a small spark of their chemical obligations, react in a car engine to produce the usual mixture of exhaust gases; the exhaust gases, so formed, then disperse into the atmosphere.

Making snowballs will always make one's hands colder, because thermal energy always flows from hot to cold. No material can spontaneously gain thermal energy from colder surroundings. We take this for granted, but this simple fact is at the heart of the nature of entropy.

We must ask ourselves why the opposite processes do not occur. Can water, *of itself*, flow upward? If we wait long enough, will the pieces of a broken plate spontaneously reorganize themselves into their previous arrangement? Can carbon monoxide, carbon dioxide, and water vapor ever congregate and react to form petroleum and oxygen? Has anything ever been known to get hot in the refrigerator? Of course, these are foolish questions because no one has yet seen these things happen. In many instances, energy is converted from one form to another, the total energy being conserved in each case (first law). Falling water represents the conversion of gravitational energy to kinetic, and then to thermal energy. (Water gains about 0.1°C in temperature when it falls 40 m.) Chemical energy is converted into thermal energy in the cylinder of the gasoline engine, although little, if any, energy is involved in the dispersion of cooled exhaust gases. In no case is the first law of thermodynamics able to say anything about the direction of these common occurrences.

4.2 EXOTHERMICITY: A POSSIBLE CRITERION

The problem of spontaneity, as outlined above, occupied many eminent scientists in the latter half of the nineteenth century. In the case of chemical reactions carried out under the usual conditions of constant temperature and pressure, it was initially assumed that if, and only if, a reaction gave out heat, it occurred spontaneously. It turns out that a second factor is involved, and that both factors must be considered. By considering a number of examples in which ΔH is negative, zero, or positive, this second factor, which is best described as a "tendency to randomness," will become more familiar.

4.2.1 SPONTANEOUS EXOTHERMIC PROCESSES

Most exothermic chemical reactions, including oxygen combustions, are spontaneous in the sense that the reactants are unstable compared with the products. The reaction may well need a catalyst in order to make it occur at an appreciable rate, or perhaps a spark may be necessary to initiate the process. The rate may be very low, but the direction of the process is fixed.

Another example of a spontaneous, exothermic process will be familiar to skiers. Fresh snow is powdery and affords smooth and easy skiing, but over a period of weeks, the conditions deteriorate. By a process of progressive cannibalization, large crystals grow at the expense of the small. For a given mass of snow, the surface area and, hence, the surface energy is greater for small crystals and so their vapor pressure is very slightly higher. The overall effect is to convert the potential surface energy to thermal energy by altering the size distribution of the ice crystals.

On the basis of arguments such as these, M. Berthelot and J. Thomson became convinced that exothermicity was the only factor involved in determining chemical affinity. We shall see that many spontaneous processes have zero, and even positive, values of ΔH, which clearly shows that this is only part of the story.

4.2.2 SPONTANEOUS PROCESSES INVOLVING NO HEAT CHANGE

(a) Suppose that the water in a covered swimming pool is allowed to come to complete rest, with no eddies or convection currents. If a drop of strong aqueous dye solution is carefully added–not dropped–at the surface, the color will quickly spread. Similarly, if a girl wearing a scent walks quietly into a large room, it is only a matter of minutes before the molecules of the scent will have diffused throughout the room, even if the air is perfectly still. (Such behavior of gases and liquids is no different for near-ideal gases or solutions. It does not depend on nonideality.) Both of these processes may be represented schematically, as shown in Figure 4.1. Infinitesimally small amounts of thermal energy are involved in such processes. Even as ideal behavior is approached, and $\Delta H = 0$, no diminution of reactivity is observed.

(b) In 1843, Joule performed a classic experiment, which is depicted in Figure 4.2. Dry air was introduced into the left container, whereas the right one was evacuated. The temperature was read carefully and noted and then tap C was opened. The gas immediately rushed through the stopcock and filled all the available space. Joule

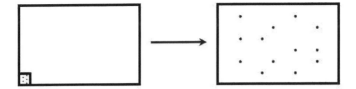

FIGURE 4.1   Process in which concentrated matter disperses.

noticed no change of temperature of the water, even with his delicate thermometer. (Joule's calorimeter was actually not sensitive enough to measure the very small temperature change involved, which is due solely to the nonideality of the gas.) Helium, which is nearly ideal under ambient conditions, behaves in exactly the same manner, although there is only a minute thermal change and it is a reasonable assumption that a truly ideal gas would expand in a similar manner.

Moreover, although the gas is turbulent when the tap is opened, no net work is done on the surroundings. This process is a natural, irreversible process and occurs even though $q = w = 0$. Thus, $\Delta U = 0$, and so the change of U with volume, under isothermal conditions, is zero. That is:

$$\left(\frac{dU}{dV} \right)_T = 0.$$

The important point to note, however, is that the process occurs, although no heat is liberated. (What general feature is common to processes (a) and (b)? The diagrams help answer this question.)

(c) A chemical reaction with $\Delta H = 0$ will now be discussed. Silver chlorite, $AgClO_2$, is unstable with respect to its elements:

$$AgClO_2(s) \rightarrow Ag(s) + \tfrac{1}{2}Cl_2(g) + O_2(g).$$

It decomposes when heated gently, and shows no tendency to re-form. However, it so happens that, quite by chance, ΔH is as near zero as can be measured. Something

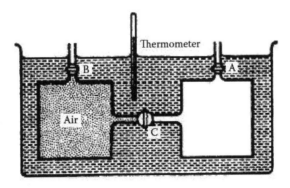

FIGURE 4.2   Joule's gas expansion experiment. On allowing gas to flow into the right-hand container, no temperature change was observed.

other than exothermicity is the driving force behind this reaction. It is significant that 1.5 mol of gas are formed per mole of decomposing solid silver chlorite.

4.2.3 ENDOTHERMIC PROCESSES

(a) If a little diethyl ether is placed in an open dish, spontaneous evaporation occurs, and this requires thermal energy to maintain a constant temperature. A cooling effect is observed. Use is made of this effect when ethyl bromide is used as a local anesthetic. Its boiling point, 38.4°C, is very close to normal body temperature, and so evaporation is very fast. The process is:

$$C_2H_5Br(l) \rightleftarrows C_2H_5Br(g), \text{ and } \Delta H^0 = 27.36 \text{ kJ.mol}^{-1}.$$

Evaporation occurs to some extent from any liquid surface. Moreover, all evaporation processes are endothermic. In common with the silver chlorite decomposition mentioned above, this process leads to a significant volume increase.

(b) There are many spontaneous but endothermic chemical processes, although few occur at laboratory temperatures. Consider the following experiment. Thirty grams of barium hydroxide and 15 g of ammonium thiocyanate are placed in an Erlenmeyer flask and the mixture is stirred with a glass rod. The flask is placed on a wet sponge or moist (wooden) board. The reaction mixture turns liquid and the sponge or board beneath the flask freezes. In addition, the smell of ammonia is perceptible. The temperature in the reaction vessel can easily fall to −20°C. The following reaction takes place:

$$Ba(OH)_2 \cdot 8H_2O(s) + 2NH_4SCN(s) \rightarrow Ba(SCN)_2(s) + 2NH_3(g) + 10H_2O(l).$$

This reaction causes significant cooling under normal laboratory conditions, and the enthalpy change for the reaction must be positive [ΔH^0 cannot be calculated because $\Delta_f H^0$ (Ba(SCN)$_2$) is not known]. In this and similar spontaneous endothermic chemical reactions, gaseous product arises from condensed reactants. We shall see later that this "gasification" is closely related to the increase in entropy or randomness associated with the reaction. It seems that when an unfavorable, positive value of ΔH is overcome in a reaction, this is associated with randomization.

4.3   THE SECOND DRIVING FORCE

We have tested the idea that exothermicity is alone responsible for spontaneous reactions, and found it severely lacking. It does, however, contain some virtue in that sufficiently large endothermicities do indicate stability of reactants. Thus, thermal decomposition of Al_2O_3, or CaO, does not occur, even at quite high temperatures and the decomposition reactions have very large positive ΔH^0 values (+1117 and +1270 kJ mol^{-1} of O_2, respectively).

That exothermicity is not the only factor is shown clearly in Section 4.2.2. Here, where $\Delta H = 0$, the changes were characterized by a spreading of at least some of the materials into a larger volume. Diffusion of the dye in the swimming pool involved

TABLE 4.1
Numerical Values of Entropy for Some Compounds at 25°C

Material	Physical State	Entropy, S^0 (J mol^{-1} K^{-1})
C	Crystal, graphite	5.69
ZnO	Crystal	43.7
Zn	Crystal	41.6
Kr	Gas	164.1
ZnSO$_4$·H$_2$O	Crystal	145.5
CO	Gas	197.7

randomization, in a simple geometric sense. At the start, we at least knew where it was, with an uncertainty of only a millimeter or so. The decomposing silver chlorite started as a crystalline solid, but the products included 1.5 mol of gas for every mole of solid silver, and this too represents an increase in randomness.

The concept of entropy is a difficult one. An understanding of it comes with repeated encounters in different contexts. Although a formal definition will be given later, it is sufficient for the time being to associate it pictorially with the degree of randomness of a material, as exemplified in the processes discussed above. It is possible to assign absolute numerical values to this entropy, S^0, which refer to individual materials in a particular physical state. In Table 4.1, values are given for a few compounds.

As a general guide, it will be found that entropy values increase during the transitions solid to liquid to gas, as demonstrated by the values for solid zinc and for gaseous krypton, at 1 atm. On the other hand, entropy will also increase with molecular complexity, as shown by ZnSO$_4$·H$_2$O(s), which has an entropy almost as great as that of gaseous krypton. The qualitative effects are nonetheless clear. The descriptive approach to entropy adopted here is adequate for the isothermal processes chosen as examples. It is not easy, for example, to visualize the change of entropy that occurs when two bodies at differing temperatures come into thermal contact. (Entropy increases.) The mathematical approach developed in the next chapter will, however, cope with this and other difficulties that arise.

PROBLEMS

4.1   Assuming that all molecules below are gaseous, predict which one has the greater entropy under the same conditions.

$$\begin{array}{ccc} H_2C & \!\!\!\!\!-\!\!\!\!\! & CH_2 \\ | & & | \\ H_2C & \!\!\!\!\!-\!\!\!\!\! & CH_2 \end{array}$$

(a) (structure above) vs. $CH_3CH_2CH_2CH_3$
(b) $CH_3CH_2CH_2CH_3$ vs. $CH_3C(CH_3)_2CH_3$
(c) N_2O vs. N_2O_4
(d) C_2H_6 vs. C_3H_8

4.2   The Sachse process (Problem 3.4) is highly endothermic. How can it ever be made to operate effectively? Use entropy values from Appendix III to develop your argument.

4.3   The example in item (c) of Section 4.2.2, the decomposition of silver chlorite, proceeds with entropy increase. Using values of absolute molar entropies (S_{298}^0) from Appendix III for the various substances, calculate ΔS^0 for the process as written.

4.4   Thionyl chloride is a very useful drying agent because the products are completely gaseous and contamination is usually negligible. It reacts with free water (and also with water of crystallization), according to the equation:

$$SOCl_2(l) + H_2O(l) \rightarrow SO_2(g) + 2HCl(g).$$

Using data from Appendix III, calculate ΔH^0 and ΔS^0 for this reaction. (Why do you think this reaction occurs spontaneously?)

4.5   The experiment described in item (b) of Section 4.2.3 can also be carried out with $NH_4Cl(s)$ instead of $NH_4SCN(s)$.

Consider the following reaction:

$$Ba(OH)_2 \cdot 8H_2O(s) + 2NH_4Cl(s) \rightarrow BaCl_2 \cdot 2H_2O(s) + 2NH_3(g) + 8H_2O(l)$$

Using data from Appendix III calculate ΔH^0 and ΔS^0 for this reaction.

5 Entropy

M. Jourdain: What? When I say, "Nicole, bring me my slippers and give me my night-cap," that's prose?
Le Maître de Philosophie: Yes, sir.
M. Jourdain: Gracious me! I've been talking prose for the last forty years and have never known it.

(Molière, Le Bourgeois Gentilhomme)

Changes in entropy are occurring all around us, but for most of us, like M. Jourdain with his prose, awareness can come quite late in life. Although we cannot detect these changes with our physical senses, it is possible to calculate them from simple measurements. After defining entropy, and performing such calculations, the concept will become progressively more familiar.

5.1 MEASUREMENT OF ENTROPY

5.1.1 THE SECOND LAW OF THERMODYNAMICS

The concept of entropy has come about slowly, with the accumulation of much human experience and is embodied in the so-called "second law," which is a statement of this practical experience, and no more. It should be stressed that exceptions to the "law" may yet be found. It may be stated in many different ways. Its origins are closely bound to the science of heat engines.

There exists a thermodynamic function of state, called entropy, S. The entropy of system and surroundings increases together during all natural or irreversible processes:

$$\Delta S_{sys} + \Delta S_{surr} > 0.$$

For a reversible process, the total entropy is unchanged:

$$\Delta S_{sys} + \Delta S_{surr} = 0.$$

Notice that entropy is not conserved; this is difficult to accept, although the evidence is in fact quite convincing.

A small change in entropy of dS is measurable in terms of the thermal energy q_{rev} added to a system during a reversible process, and the (constant) temperature T, at which it is performed:

$$dS = \frac{dq_{rev}}{T}. \tag{5.1}$$

Because S is a function of state, the term dq in Equation 5.1 must be quite unambiguous. In fact, as we shall see, reversibility necessarily implies that maximum work is involved, and that q is specified.

Because no violations of the second law have ever been observed, there is no reason to doubt its universal applicability. However, in June 1993, there was uproar in the scientific community among physicists and chemists. Professor Silbergleit from the Ioffe Institute of St. Petersburg in Russia stated that he could develop a heat engine in which heat could be turned nearly completely into work, with negligible losses. Such an engine would be a perpetual motion machine. One of the formulations of the second law of thermodynamics, however, is that if heat turns into work, considerable losses will occur in a cyclic process. Professor Silbergleit's assertions were subsequently discredited in a meeting at The Netherlands Energy Research Foundation ECN in Petten in June 1993.

5.1.2 REVERSIBILITY AND ENTROPY

In general, energy or "work" is the product of an *intensive factor* and an *extensive factor*. Intensive factors are those properties that are independent of the size of the system, such as temperature, pressure, odor, and density. An extensive factor, on the other hand, is one that depends on the size of the system, such as mass, volume, and internal energy.

An easy way to determine whether a factor is intensive or extensive is to divide the system into two equal parts with a partition. Each part will have the same value of intensive factors as the original system, but half the value of the extensive factors.

$$\text{Mechanical energy} = \text{force} \times \text{distance},$$

$$\text{Work of compression} = \text{pressure} \times \text{volume change},$$

$$\text{Electrical energy} = \text{voltage} \times \text{quantity of charge}.$$

In each case, if the intensive factor (force, pressure, or voltage) is opposed by a force, pressure, or voltage that is *virtually equal* to it, then energy is transferred *reversibly*. The process will occur infinitely slowly, but will result in the transfer of maximum energy (maximum work). The eventual winners of a tug-of-war between two evenly matched sides will be working "flat-out" for their win. The work done by the winning team depends on the resistance they meet, not on their own inherent strength. The maximum work will be done when they are virtually equal to their opponents, but the performance will then take a long time. They are working reversibly because a minute increase of effort by their opponents will reverse the direction of the change.

It is this concept of reversibility that must be applied when entropy changes are calculated. There is, however, no reason why the actual process should occur reversibly, because entropy is a function of state, and ΔS is the same, whatever the route. It is frequently necessary to devise, on paper, a reversible route merely to perform the entropy-change calculation. Let us apply these ideas to specific examples.

5.1.2.1 Isothermal Expansion of Gases

The gas expansion experiment of Joule, shown in Figure 4.2, took place completely irreversibly; no work was done, because no resistance was offered from outside. The overall change is:

$$Gas(T,V_1) \rightarrow Gas(T,V_2).$$

Imagine a reversible path for this change in which an opposing piston offers virtually equal, but opposite, pressure. As infinitesimal expansion occurs, so that the opposing pressure drops by an infinitesimal amount, and expansion continues, albeit infinitely slowly. Negative work is being done on the gaseous system by the piston (surroundings), and to maintain (constant) temperature, heat must be added. Internal energy is then kept constant and so for a small change:

$$dU = 0 = dq_{rev} + w = dq_{rev} - PdV \text{ and so } dq_{rev} = PdV.$$

On integrating, and making the ideal gas assumption:

$$\int dq_{rev} = \int P\,dV = nRT \int \frac{dV}{V}.$$

$$q_{rev} = nRT \ln V_2/V_1 \quad (T \text{ constant}). \tag{5.2}$$

From Equation 5.1, for a finite change:

$$\Delta S = \frac{q_{rev}}{T} = nR \ln V_2/V_1 \quad (T \text{ constant}). \tag{5.3}$$

The entropy change for the irreversible Joule experiment may thus be calculated *from the hypothetical, reversible process*, because S is a function of state only.

5.1.2.2 Reversible Transfer of Heat

Consider a countercurrent heat exchanger, such as is used in many industrial plants. In essence, this is a pair of pipes in close thermal contact (Figure 5.1). If it were to be operated reversibly, the water and initially hot liquid would be in continuous equilibrium with, at each point, only an infinitesimal thermal gradient between the liquids. In such a case, the liquids are virtually stationary, and maximum energy will be transferred. Entropy changes at point A are:

$$dS_{liquid} = -\frac{dq}{T_{liquid}}, \; dS_{water} = +\frac{dq}{T_{water}},$$

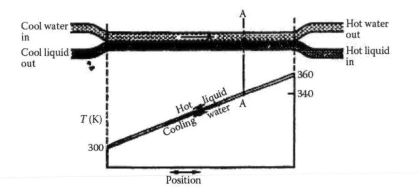

FIGURE 5.1 Simplified heat exchange. Reversible operation requires infinitely slow liquid movement; the outgoing liquid then has the same temperature as the inlet water.

and the total entropy change for liquid and water ("system and surroundings") is zero:

$$dS_{total} = dq\left(-\frac{1}{T_{liquid}} + \frac{1}{T_{water}}\right) = 0. \qquad (5.4)$$

On the other hand, suppose that the liquids move at a finite speed. Now, T_{liquid} is greater than T_{water}, and Equation 5.4 gives $dS_{total} > 0$. Entropy has increased because, and to the extent that, the process is irreversible.

Example 5.1

One mole of water at 0°C freezes to ice in a well-insulated vessel containing kerosene at a temperature of 0°C:

$$\Delta_{fus}H = 6.008 \text{ kJ mol}^{-1}.$$

Calculate the entropy change of the water, the kerosene, and the total entropy change within the vessel (ΔS_{sys}).
    In this case, the path chosen for computation is quite simple because the process described above is a reversible one:

$$1 \text{ mol } H_2O(l) \underset{\text{reversible freezing}}{\overset{\Delta S_{water}}{\rightleftharpoons}} 1 \text{ mol } H_2O(s)$$
$$\phantom{1 \text{ mol } H_2O(l)} 273.15 \text{ K} \qquad\qquad\qquad 273.15 \text{ K}$$

$$\Delta S_{water} = \frac{\Delta_f^s H}{273.15} = \frac{-6008}{273.15} = -22.00 \text{ J K}^{-1}.$$

The entropy change experienced by the water is negative, which is not surprising because it turns completely from liquid into solid at the same temperature (Section

5.2). However, heat is released *to* the kerosene, which in turn must have a positive change in entropy. The heat given to the kerosene is

$$\Delta S_{\text{ke rosine}} = \frac{+6008}{273.15} = +22.00 \text{ J K}^{-1}$$

$$\Delta S_{\text{sys}} = \Delta S_{\text{water}} + \Delta S_{\text{kerosene}} = -22.00 + 22.00 = 0 \text{ J K}^{-1}.$$

Because the process is adiabatic, the surroundings undergo no change as a result of this process and, hence, $\Delta S_{\text{surr}} = 0$, which is in accordance with the second law of thermodynamics, which states that the total change in entropy for a reversible process is zero.

5.1.2.3 An Irreversible Change

After heat treatment, a steel bar at 1000 K is quenched in warm water (333 K). This is a typical irreversible process. Consider the transfer of the first joule, which leaves the steel at 1000 K, and arrives in the water at 333 K. For purposes of calculation, this process could be considered in two reversible parts. First a joule is removed from the steel by contact with a thermostat set at $(1000 - dT)$ K; then a joule is added to the water from a bath at $(333 + dT)$ K.

The entropy changes are, for the reversible process:

$$\Delta S_{\text{steel}} = -\frac{1}{1000}, \quad \Delta S_{\text{hot bath}} = +\frac{1}{1000}$$

$$\Delta S_{\text{water}} = +\frac{1}{333}, \quad \Delta S_{\text{cold bath}} = -\frac{1}{333}$$

$$\Delta S_{\text{total}} = 0 \quad \text{(reversible)}.$$

In the irreversible quenching, only steel and water are involved, for which the entropy changes are already calculated, so here:

$$\Delta S_{\text{total}} = -\frac{1}{1000} + \frac{1}{333} = +0.002 \text{ J K}^{-1} \quad \text{(irreversible)}.$$

Once again, entropy has increased during the irreversible process.

Example 5.2

One mole of supercooled liquid water at −5°C freezes to ice in a well-insulated vessel containing kerosene at a temperature of −5°C:

$C_p(\text{ice}) = 37.8$ J mol^{-1} K^{-1}, $C_p(\text{water}) = 75.3$ J mol^{-1} K^{-1}, $\Delta_{\text{fus}}H = 6.008$ kJ mol^{-1}

Calculate the entropy change of the water, the kerosene, and the total entropy change within the vessel.

Because the entropy is a function of state, its change between any two states is independent of path, and the path chosen for the computation is a reversible one (Figure 5.2):

$$\Delta S_{water} = \Delta S_1 + \Delta S_2 + \Delta S_3 = \int_{286.15}^{273.15} \frac{C_p(l)}{T}\, dT + \frac{\Delta_f^s H}{273.15} + \int_{273.15}^{268.15} \frac{C_p(s)}{T}\, dT$$

$$\Delta S_{water} = 75.3\ln\frac{273.15}{268.15} - \frac{6008}{273.15} + 37.8\ln\frac{268.15}{273.15} = -21.30 \text{ J K}^{-1}.$$

As in Example 5.1:

$$\Delta S_{kerosene} = \frac{\left|\Delta_f^s H\right|}{268.15} = \frac{\left|\int_{268.15}^{273.15} C_p(l)dT + \Delta_f^s H + \int_{273.15}^{268.15} C_p(s)dT\right|}{268.15}$$

$$\Delta S_{kerosene} = \frac{\left|(75.3\times 5) - 6008 + (37.8\times -5)\right|}{268.15} = +21.71 \text{ J K}^{-1}$$

$$\Delta S_{sys} = \Delta S_{water} + \Delta S_{kerosene} = -21.30 + 21.71 = +0.41 \text{ J K}^{-1}.$$

So within the insulated vessel the total entropy change is positive, in accordance with the second law of thermodynamics, because the freezing of supercooled water at −5°C is a spontaneous process.

(As in Example 5.1, $\Delta S_{surr} = 0$, because the process is adiabatic.)

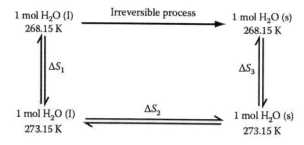

FIGURE 5.2 Entropy change for the irreversible process is equal to the sum of entropy changes for processes 1, 2, and 3.

5.1.3 CHANGES IN ENTROPY WITH TEMPERATURE

So far, we have considered only isothermal processes. How do we calculate the entropy change for a gas that is heated? Now, $dS = dq_{rev}/T$ and $C_P = dq/dT$; therefore, as dq and dq_{rev} are identical:

$$dS = \frac{C_P}{T} dT$$

and

$$\Delta S = S_2 - S_1 = \int \frac{C_P}{T} dT \quad (P \text{ constant}). \tag{5.5}$$

Over the temperature range, where C_P may be considered constant:

$$S_2 - S_1 = C_P \int \frac{1}{T} dT = C_P \ln T_2/T_1. \tag{5.6}$$

In general, however, it is necessary to plot C_P/T against temperature. Equation 5.5 then tells us that the area under the curve between the two temperatures of interest gives the change in entropy.

Example 5.3

The average value of C_P for silane, SiH_4, between 340 and 360 K is 47.03 J mol^{-1} K^{-1}. By how much does the entropy of 4 mol silane increase when it is heated over this temperature range?

Using Equation 5.6:

$$S_{360} - S_{340} = 4(47.03)\ln 360/340$$

$$= +10.8 \text{ J K}^{-1}.$$

For processes carried out at constant volume, an exactly analogous derivation using C_V may be followed to give:

$$\Delta S = S_2 - S_1 = C_V \ln T_2/T_1 \quad (V \text{ constant}).$$

Again, a mean value of C_V is used.

5.1.4 AN ADIABATIC COMPRESSION

If a gas is effectively insulated during reversible compression, $q_{rev} = 0$, and so $\Delta S = 0$. This means that $\Delta U = w$, and the temperature rises. Does this not mean that the entropy also increases? The answer is no, because the effect of the temperature increase is exactly matched by the reduction in volume that accompanies it.

5.2   ABSOLUTE ENTROPIES

So far, we have been concerned solely with *changes* in entropy. Is it possible to assign a zero entropy to a particular sample or state? In fact, this may be done and we can, as a result, calculate *absolute values of entropy* for individual compounds.

5.2.1   THE THIRD LAW OF THERMODYNAMICS

The best organized physical state of a compound is that of a single perfect crystal, because each atomic or molecular unit shares a definite spatial relationship with all the others. Moreover, at low temperatures, there are fewer populated energy levels, and there is correspondingly less uncertainty about the energy of any particular unit. Translational energy is lost on crystallization, rotational energy does not persist for long, and, finally, at a sufficiently low temperature, only vibrational motions remain. At absolute zero, it is generally accepted that only the ground or lowest vibrational level would be occupied. All units are vibrating, but they each have the smallest possible, or "zero point," energy. The third law may be expressed: *All truly perfect crystals at absolute zero temperature have zero entropy.*

There are, however, several ways in which a crystal may fail to be truly perfect. Isotopic mixtures of, say ^{35}Cl and ^{37}Cl give rise to an entropy of mixing; different combinations of nuclear spin, as occur in *ortho-* and *para*-hydrogen, will cause some randomization. Crystals are seldom perfect, and there are other effects that must be accounted for.

Let us now calculate an entropy in absolute terms. Equation 5.5 may be written as:

$$S_{T_2} - S_{T_1} = \int_{T_1}^{T_2} \frac{C_P}{T} dT. \qquad (5.7)$$

If $T_1 = 0$ K,

$$S_{T_2} - S_0 = \int_0^{T_2} \frac{C_P}{T} dT.$$

In general,

$$S_T = \int_0^T \frac{C_P}{T} dT. \qquad (5.8)$$

Values of C_P may be measured down to a few degrees Kelvin, but a means of extrapolation must be found for the last few degrees. The Debye theory of the heat capacity of solids gives, at low temperatures:

$$C_P = \alpha T^3 + \beta T$$

or

$$\frac{C_P}{T} = \alpha T^2 + \beta,$$

and so values of C_P/T may be extrapolated to zero by plotting against T^2. It is then possible to carry out a graphical integration of Equation 5.8 from 0 K, and determine the absolute entropy.

Example 5.4

E. F. Westrum Jr. has published low-temperature heat capacities for uranium disulfide, US_2. Use these values to determine the absolute molar entropy at 298 K. The data are shown in Table 5.1, together with the calculated values of C_P/T, and the graph of C_P/T versus temperature is shown in Figure 5.3.

On measuring the shaded area underneath the curve, using square-counting or paper-weighing techniques, the absolute entropy of US_2 at 298 K is found to be 110.4 J mol^{-1} K^{-1}. This result can be well verified by the reader using the data provided.

TABLE 5.1
Heat Capacity Data for Uranium Disulfide

Temperature (K)	C_P (J mol^{-1} K^{-1})	C_P/T (J mol^{-1} K^{-2})
8.2	0.29	0.035
9.6	0.67	0.071
11.8	1.55	0.130
14.3	3.26	0.226
17.5	5.31	0.305
21.2	8.08	0.381
26.4	11.6	0.439
33	15.5	0.469
41	20.1	0.490
51	25.1	0.494
69	33.9	0.490
94	45.2	0.481
116	53.1	0.460
142	59.8	0.423
171	64.9	0.381
207	69.0	0.335
244	71.8	0.293
279	73.6	0.264
312	74.9	0.238

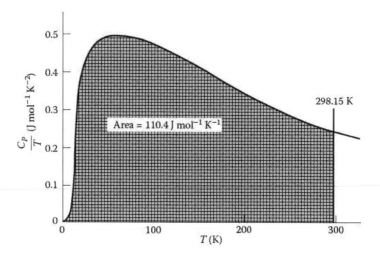

FIGURE 5.3   Heat capacity and temperature graph for determination of absolute entropy of uranium disulfide.

5.2.2 ΔS FOR PHASE CHANGES

Uranium disulfide is a particularly simple case, because it is solid throughout the experiment, and exists in only one crystalline form. Many compounds, however, undergo solid modifications or phase changes. These changes are easily catered to, because they occur at fixed temperatures.

Thus, for fusion:

$$\Delta_{fus}S = \int \frac{dq_{rev}}{T_{fus}} = \frac{\Delta_{fus}H}{T_{fus}} \qquad (T, P \text{ constant}).$$

In general, for any phase change,

$$\Delta S = \frac{\Delta H}{T} \qquad (T, P \text{ constant}). \tag{5.9}$$

Using the third law, absolute entropies of any substance in its standard state, at 298 K, written S_{298}^0, may be determined from thermal measurements, using Equations 5.6 (heat capacities) and 5.9 (enthalpies and temperatures of transition) as appropriate.

Example 5.5

Data are available for isobutane, from 12.53 K upward. It does not undergo any solid/solid transitions, but melts at 113.74 K ($\Delta H = 4541$ J mol^{-1}) and boils at 261.44 K ($\Delta H = 21.29$ kJ mol^{-1}).

The different stages in the calculation of S^0_{298} are summarized below.

$$\Delta S, \text{ J mol K}^{-1}$$

(a) $\int \frac{C_p}{T} dT$ for sold, $0 \text{ K} - T_f$,

    including Debye extrapolation below 12.53 K   68.46

(b) Fusion, $\Delta H/T = 4541/113.74$ 39.93

(c) $\int \frac{C_p}{T} dT$ for liquid, $T_f - T_b$ 92.68

(d) Boiling, $\Delta H/T = 21,290/261.44$ 81.45

(e) $\int \frac{C_p}{T} dT$ for gas, $T_b - 298.15 \text{ K}$ <u>12.17</u>

$$S^0_{298} = 294.7 \text{ J mol}^{-1} \text{ K}^{-1}$$

It is helpful to represent the entropy changes of isobutane pictorially, as in Figure 5.4. Processes (a) and (c) increase the temperature, and the entropy increases because there is a wider range of possible molecular energy states available. Processes (b) and (d) represent increases in geometrical randomness.

Figure 5.5 shows in detail how the entropies of hydrogen and of boron increase with temperature. Notice particularly how great the contribution due to vaporization is. Crystalline boron is very highly ordered, and has correspondingly low entropy.

Values of S^0_{298} are obtained from heat capacity and other measurements, and Appendix III includes a selection of such data for a number of compounds. From such tabulations, it is possible to determine values of S^0_{298} for any specified reaction. As usual, one can use an equation similar to Equation (1.1):

$$\Delta S^0_{298} = \sum S^0_{298}(\text{Products}) - \sum S^0_{298}(\text{reactants}). \tag{5.10}$$

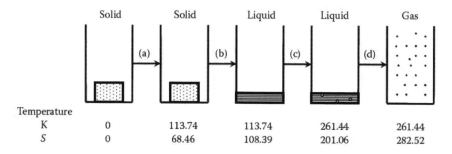

	Solid	Solid	Liquid	Liquid	Gas
Temperature		(a)	(b)	(c)	(d)
K	0	113.74	113.74	261.44	261.44
S	0	68.46	108.39	201.06	282.52

FIGURE 5.4 Entropy of 1 mol isobutane, C_4H_{10}, is represented pictorially. The crystal at 0 K is assumed to be perfect in every respect. Processes (a) and (c) represent temperature increases, processes (c) and (d) indicate phase changes, and the entropy rises during each.

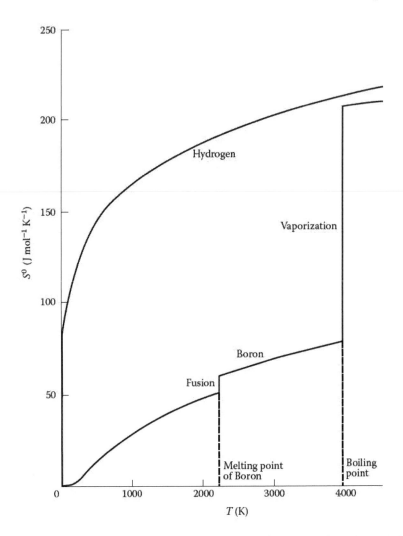

FIGURE 5.5 The standard molar entropies of boron and hydrogen from 0 to 4500 K. Note the large increase in vaporization. (Data from *JANAF Thermochemical Tables*, 1965. Dow Chemical Company, Midland, MI. With permission.)

Example 5.6

What is ΔS^0_{298} for the formation of isobutane from its elements? The reaction is:

$$4C(\text{graphite}) + 5H_2(g) \rightarrow C_4H_{10}(g).$$

Using values from Example 5.5 and Appendix III, we find:

$$\Delta S^0_{298} = 294.7 - 4(5.7) - 5(130.7)$$

$$= -381.6 \text{ J mol}^{-1} \text{ K}^{-1}.$$

5.3 THE DIRECTION OF TIME

It has been said that increasing entropy is the arrow of time; but what is the connection between them? All irreversible, natural processes lead to an increase in entropy with respect to time. If we build a large house of cards, this represents a very unlikely assemblage; there is only one house arrangement but a multitude of disorganized arrays. All too easily, a small vibration causes reversion to a disordered state. This occurrence may cause minor dismay, but on the other hand to watch the cards spontaneously jumping and maneuvering themselves into a perfect structure would suggest either mental derangement or that time was going backwards.

Left on its own, a bouncing ball will eventually come to rest. A small ball bearing dropped onto a sheet of plate glass will rebound almost to its previous height each time, and the various changes are almost reversible. Eventually, however, the kinetic (or gravitational or distortional) energy will be converted to thermal energy, which is simply a randomized kinetic energy operating on a molecular scale. The motionless ball will be a little hotter than before; the molecules will be jiggling all in different directions with varying energies.

That all these molecules should spontaneously organize themselves into coordinated upward movement, so that the ball jumps up once again, is not impossible in terms of energy conservation. It is, however, extremely unlikely, because there is only one type of net upward movement compared with the myriads of equally likely schemes of random movement.

The vast number of possible schemes of random movement is related closely to the greater entropy of the disordered condition. For all natural processes, the total entropy of the system plus surroundings always increases with increasing time; entropy is constantly produced. If a process or change is imagined for which entropy is destroyed, then the direction of time must be reversed for that process, but this cannot happen. A given system, in given surroundings, always changes in the same direction.

REFERENCE

1. Smith, W. T., G. D. Oliver, and J. W. Cobble. 1953. Thermodynamic properties of technetium and rhenium compounds. *J Am Chem Soc* 75:5785–5786.

PROBLEMS

5.1 Decide by inspection whether the following reactions will have values of ΔS that are positive, negative, or about zero. Check values from Appendix III.

$$2H_2(g) + C_2H_2(g) \rightarrow C_2H_6(g)$$

$$LiH(s) + H_2O(l) \rightarrow LiOH(s) + H_2(g)$$

$$C(graphite) + CO_2(g) \rightarrow 2CO(g)$$

$$NO(g) + O_3(g) \rightarrow NO_2(g) + O_2(g)$$

$$C_2H_4(g) + Br_2(l) \rightarrow CH_2BrCH_2Br(l).$$

5.2   A 2-kW electric heater runs for an hour-and-a-half in a room at 20°C.
The filament temperature is 800°C. What is the overall increase in
entropy in J K^{-1}?

5.3   It is possible to cool a liquid below its freezing point without the formation of a solid, if care is taken to prevent nucleation. For example, 39 g
of liquid benzene with a freezing point of 5.4°C are in contact with
a cooling system and are cooled slowly from 45°C to −10°C without
freezing. At −10°C the subcooled benzene spontaneously crystallizes
by introduction of a small crystal of benzene. After the crystallization
is complete, the benzene is cooled further to −40°C. At 5.4°C, $\Delta_{fus}H =$
126.75 J g^{-1}, $C_p(l) = 1.673$ J g^{-1} K^{-1} and $C_p(s) = 1.255$ J g^{-1} K^{-1}. Devise
a reversible route to calculate $\Delta S_{benzene}$. If the benzene together with
the cooling system were placed into a well-insulated vessel, explain
whether ΔS_{tot} in the vessel is positive, negative, or zero.

5.4   A leaden ball having a mass of 1 kg and a temperature of 227°C is
placed in a well-insulated vessel containing 5 kg of water initially at
27°C. The heat capacities are 4.185 J g^{-1} K^{-1} for water and 0.13 J g^{-1}
K^{-1} for the leaden ball. Find the temperature of this system in the final
state by setting up a heat balance. (Hint: Be careful about the sign on
the left- and right-hand site of this equation.) Calculate the entropy
change of the ball, the water, and the total process within the vessel.

5.5   Values of C_p for silane increase steadily with temperature:

T (K)	200	250	300	350	400
C_p (J mol^{-1} K^{-1})	35.5	39.2	43.0	47.4	51.5

Determine the increase in entropy ($S_{400} - S_{250}$) on heating 1 mol silane
from 250 to 400 K at 1 atmosphere; use a graphical or a numerical
technique with a computer program.

5.6   Smith et al. [1] have published low-temperature heat capacity data for
metallic rhenium.

T (K)	C_p (J mol^{-1} K^{-1})	T (K)	C_p (J mol^{-1} K^{-1})
20.39	0.661	80.05	14.91
24.89	1.318	92.88	17.23
31.74	2.61	103.04	18.39
34.66	3.35	121.82	20.43
39.98	4.540	148.41	22.27
48.89	7.544	202.65	24.28
62.26	11.42	244.81	25.45
70.82	13.12	300.01	25.73

Determine S^0_{298} (Re) graphically, using a $C_p = A.T^3$ extrapolation at
very low temperatures, where A is a constant, or by means of numerical method with a computer program. (Calculate constant A from the
value of C_p at 20.39 K.)

5.7 From HSC Chemistry (Reference 1, Chapter 6), we find the following equation for the heat capacity for hydrogen between 298.15 and 5000 K:

$$C_p(H_2) = 25.8744 + 4.8399 \times 10^{-3}T + 1.5868 \times 10^5 T^{-2} - 3.726 \times 10^{-7}T^2.$$

Calculate the increase in entropy on heating 1 mol hydrogen from 298.15 to 600 K.

5.8 The hexafluorides of tungsten, osmium, and iridium all undergo solid-solid transitions at, or a little below, 0°C. The process β (low-temperature form) → α (high-temperature form) is endothermic in each case:

	WF$_6$	OsF$_6$	IrF$_6$
$\Delta_{trs}H$ (kJ mol^{-1})	5.86	8.24	7.11
T (K)	265.0	272.8	273.6

Calculate the increase of entropy for these transitions.

5.9 Calculate ΔS^0 for the decomposition of calcium carbonate into carbon dioxide and calcium oxide at 1 bar:

$$CaCO_3(s) \rightarrow CaO(s) + CO_2(g)$$

at 25°C and 600°C. $C_p(CaCO_3) = 81.88$ J mol^{-1} K^{-1}, C_p (CaO) = 42.80 J mol^{-1} K^{-1} and $C_p(CO_2) = 37.12$ J mol^{-1} K^{-1}.

6 Free Energy: The Arbiter

In this chapter, we come to the most important single concept in chemical thermodynamics, that of Gibbs free energy, G. It is a function of state that provides the criterion for deciding whether a change of any kind will tend to occur. Physical and chemical changes will have simple numerical values of ΔG such that a negative number indicates spontaneity, a positive number tells us that the reverse change tends to occur, and a value of zero implies an equilibrium situation.

6.1 PROCESSES IN ISOLATED SYSTEMS

We recall that whenever a process occurs in a natural or everyday situation, it does so irreversibly, and the overall change of entropy of the system and surroundings, together, is positive. This, then, is a general criterion of the spontaneity or, otherwise, of a change. It is very difficult, however, to carry out experiments in which changes occurring both in the reaction vessel (the system) and in the surroundings are followed, because the surroundings are so ill-defined. Progress is often made when a problem is reduced to its simplest form, and here it is helpful to isolate the experiment to leave the surroundings unchanged. First, suppose we keep the total volume constant and use a thermally insulated vessel. This was already applied in Chapter 5 in Examples 5.1 and 5.2. (Afterwards, we can extend our activities to include constant pressure processes.) Consider the terms ΔU, q_{to} and w_{on} under these conditions of isolation. Insulation implies that $q = 0$, and constant volume requires that the work done by expansion is also zero. Therefore $\Delta U = q + w$ must itself be zero.

First consider a small (60 mL) container of 0.03 mol of gas under high pressure suspended in a 10-L evacuated vessel, which is vacuum insulated. The little bottle breaks on falling, and the space above it is immediately filled with gas. There is no temperature change, but the gas is obviously much more spread out. To determine by how much the entropy has increased, we must imagine the same overall change to be carried out reversibly, as described in Section 5.1.2.1.

By applying Equation (5.3), $\Delta S = nR \ln V_2/V_1$, we have in this case:

$$\Delta S = 0.03(8.314)\ln\frac{10,000}{60} = 1.26 \text{ J K}^{-1}.$$

As a second example, we shall consider another experiment in which the process is quite spontaneous, and irreversible. It results, therefore, in an entropy increase, but the visual concept of randomization is difficult to apply. It is possible to relate, mathematically, the entropy of a material to its degree of disorder. This involves a detailed analysis of the distribution of all the atoms and molecules in the various energy levels available to the system. For any particular set of conditions (temperature, volume, and so on) or macrostate, there is a large number of compatible microstates. This number is denoted by ω, the so-called multiplicity of the macrostate.

A saturated hot solution of sodium thiosulfate is cooled slowly, and, if care has been taken to keep the materials clean and dust-free, supercooling by many degrees is possible. The solution is then sealed in a Dewar flask. On dropping a tiny seeding crystal through a hole in the lid (Figure 6.1) crystallization occurs, with an apparent increase of organization. This story tells us very forcefully that visual disorganization and entropy are not synonymous, for although highly regular crystals have been formed, the spontaneity of the process requires increased entropy. This means that the number of equivalent arrangements of microstates must be higher than before. Nevertheless, those associated with position appear to have decreased in number due to the clustering of thiosulfate and sodium ions into crystals. The volume change is negligible and cannot account for this apparent anomaly. What other possibilities are there? The missing factor is, of course, the temperature, which increases. (ΔU for the system can still be zero, because new materials are formed, and each has its own contribution to make to the total internal energy.) This makes available a greater spectrum of thermal microstates, and so the entropy increases after all.

To summarize,

$$\Delta U = q + w = 0 \qquad \text{(System isolated)},$$

$$\Delta S_{surr} = 0 \qquad \text{(System isolated)},$$

$$\Delta S_{system} \text{ is positive} \qquad \text{(Irreversible process,}$$

$$\text{and system isolated).}$$

In both of the foregoing examples, we have seen that an increase in entropy is a valid criterion for the spontaneity of a process. In addition, the second example shows that it is often misleading to interpret entropy simply as "mixed-upness." When we move over to normal laboratory conditions, we will find it even more difficult to keep

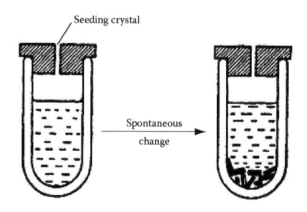

FIGURE 6.1   Spontaneous organization? Precipitation of sodium thiosulfate from a super-saturated solution. Entropy, in fact, increases because of a temperature rise.

track of what is going on. The introduction of Gibbs free energy, which simplifies the situation, will be most welcome.

6.2   GIBBS FREE ENERGY, G

In our search for a criterion of whether a particular change would occur, we have found that, in some cases at least, the attainment of minimum energy seems to be a useful indication. The demolition of a factory chimney always leads to the same predictable result, the attainment of minimum gravitational energy. Similarly a soap bubble will quickly adjust itself to a spherical shape so as to have a minimum surface area and, hence, minimum surface energy. The exothermic nature of a reaction, where energy is given out as heat and the materials achieve lower energy, seems to be only a rough guide at the best of times. We do know that reactions that are strongly exothermic give very stable products, such as the formation from their elements of Al_2O_3 ($\Delta_f H^0_{298} = -1676$ kJ mol^{-1}), Fe_2O_3($\Delta_f H^0_{298} = -823.4$ kJ mol^{-1}), and $CaCO_3$ (calcite, $\Delta_f H^0_{298} = -1207$ kJ mol^{-1}). Several exceptions to this rule of thumb for processes of small exothermicity were mentioned in Chapter 4. In these cases it was shown that at constant temperature, the increase in entropy of the system seemed to be playing a dominant part. We shall see how this factor also can be assessed.

Under isothermal conditions, these two driving forces, energy and entropy, act sometimes in concert, sometimes at variance with one another. It is nonetheless possible to define a new quantity, Gibbs free energy, which enables us to arbitrate between the sometimes conflicting factors of energy seeking a minimum, on the one hand, and entropy seeking a maximum, on the other. Gibbs free energy is defined as:

$$G = H - TS. \tag{6.1}$$

It should be noted that H, T, and S are all properties of state. This means that G must also be a property of state. To appreciate the usefulness of Gibbs free energy, we shall now fully differentiate this equation, to give:

$$dG = dH - TdS - SdT.$$

The majority of processes take place at constant temperature, and under these conditions, $dT = 0$. Thus,

$$dG = dH - TdS \quad (T \text{ constant}) \tag{6.2}$$

If the pressure is constant (usually 1 atm), we may write dq for dH, giving:

$$dG = dq - TdS \quad (T, P \text{ constant}).$$

We are now in a position to appreciate how G may be used as a criterion for spontaneity. We need to bring in the second law statement, which may be summarized as:

$$dS = dq_{rev}/T$$

$$dS > dq_{irrev}/T$$

Thus, TdS, being a property of state, will always be equal to dq_{rev}. If our process is performed reversibly, that is, in a continuous state of equilibrium, then $dG = dq_{rev} - dq_{rev} = 0$. On the other hand, a spontaneous, and therefore irreversible process will have $dq_{irrev} < TdS$, and therefore $dq_{irrev} < dq_{rev}$. Then, $dG = dq_{irrev} - dq_{rev}$, and dG is a negative quantity. This is our objective. Gibbs free energy, G, decreases as long as the process is spontaneous (dG is negative). As soon as a state of equilibrium is reached, G comes to the bottom of the hill and reaches a minimum value ($dG = 0$), which specifies that no further reaction can take place.

For constant temperature processes, Equation 6.1 may be used to apply to the initial state (reactants) and the final state (products), so that:

$$G_{products} = H_{products} - TS_{products}$$

$$G_{reactants} = H_{reactants} - TS_{reactants}$$

and so

$$\Delta G = \Delta H - T\Delta S \quad (T \text{ constant}). \tag{6.3}$$

The delta has the usual significance of "products minus reactants."

If we use this new quantity, the Gibbs free energy *change*, ΔG, for a process, we are relieved of the difficulty of following changes in the surroundings. Values of ΔH and ΔS for the process are sufficient to tell us if the process is spontaneous. If ΔG is a negative number, the process will tend to occur; if zero, equilibrium is attained; and if positive, the reverse process is spontaneous. Remember that spontaneity implies only a tendency to occur, and that the rate of reaction may well be undetectably slow. This is because thermodynamics is concerned only with the initial and final states, whereas questions of collision rates and activation energy are the province of chemical kinetics. Thus, ΔG for the reaction:

$$H_2O_2(l) \rightarrow H_2O(l) + \tfrac{1}{2}O_2(g)$$

is devastatingly negative ($\Delta G = -116.7$ kJ mol^{-1}), but hydrogen peroxide may nonetheless be kept, if pure, for long periods. The presence of a catalyst, such as platinum gauze, will quickly resolve any doubts about its thermodynamic instability.

Before we discuss some specific applications of Gibbs free energy, it is necessary to interpret ΔG as the greatest useful work available from a process at constant pressure. This will be helpful later, in Chapter 9 on electrochemistry, for example.

6.3   GIBBS FREE ENERGY AND MAXIMUM WORK

Useful work can have many interpretations, depending on the circumstances. Here, it is taken to mean any work other than the work of gaseous expansion, which is an inconvenient form. (Expansion against atmospheric pressure, $P\Delta V$, often occurs, however, and must be accounted for.) Electrical work (w_{elec}) is probably the most convenient. From the first law,

$$\Delta U = q_{to} + w_{on}.$$

Ignoring all forms of work other than those already mentioned:

$$\Delta U = q - P\Delta V + w_{elec}.$$

Thus,

$$w_{elec} = \Delta U + P\Delta V - q$$

$$= \Delta H - q \ (P \text{ constant}).$$

We should note here that, although ΔH depends only on the difference between initial and final states of the system, q will vary with the reversibility or otherwise of the process.

First, if reversibly done, the second law tells us that $dq_{rev} = TdS$, and that, if temperature is constant, this may be scaled up to give:

$$q_{rev} = T\Delta S.$$

Thus,

$$w_{elec} = \Delta H - T\Delta S,$$

but from Equation 6.3

$$w_{elec, rev} = \Delta G.$$

Thus, the more negative ΔG is, the greater the useful work (electrical work, in this case) obtainable from the process will be.

If the process is *irreversible*, $q_{irrev} < T\Delta S$, and therefore $\Delta H - q$, equal to w_{elec}, will have a smaller value. Thus, for a spontaneous change, w_{elec} will be less than the best available, which is ΔG. We will see later that the maximum electrical work available from an electrical cell will be obtained under reversible conditions, where the cell e.m.f. is opposed by an infinitesimally smaller potential. The electrons are made to work their passage around the external circuit to the maximum of their ability. Under these conditions, the electrical work depends on the equilibrium voltage, E, and on the number of electrons made to go through the circuit, corresponding to nF coulombs; F is a unit of charge, the Faraday $= 96,485$ coulombs mol^{-1} of electrons and $n =$ number of moles of electrons or equivalents. This is expressed as:

$$w_{elec, rev} = -nFE \text{ joules.}$$

(The mole is used here in its most general form, as the Avogadro's number of particles.)

6.4 SOME PROCESSES IN TERMS OF GIBBS FREE ENERGY

We have now come to the stage where we can analyze phenomena occurring daily around us, in terms of the two directing influences that operate under isothermal conditions. We shall see that the two are not always acting in the same sense, and in such cases it is necessary to use Equation 6.3 to decide whether occurrence is possible. Before going further, some general comments are in order.

First, we must find numerical values for ΔH and ΔS. The enthalpy change will be derivable from the enthalpies of formation of the participants from Appendix III. Values of S^0_{298}, from the same appendix, may be similarly combined as outlined in Section 5.22, to give ΔS^0.

Second, we should note from Equation 6.3 that, whatever the sign and magnitude of ΔS may be, it will have increasing importance at high temperatures. Thus, for an isothermal process carried out at low temperature, the ΔH term is dominant. At absolute zero, both ΔS and T are zero, and so $\Delta G = \Delta H$. On the other hand, many stellar processes are dominated by entropy considerations alone. We on earth, however, are neither in the freezer nor in the fire, but somewhere between the two. As a result, we must learn to cope with both factors.

6.4.1 Adsorption Processes

Clean surfaces often take up gaseous material, which is *adsorbed* onto the surface of the solid. For example, many metals adsorb oxygen, hydrogen, and water vapor, and charcoal is well known for its adsorbent properties. Heterogeneous catalysis involves adsorption of the reacting materials, in which case modification of the adsorbed species occurs. This makes further reaction that much more likely, speeding up the overall reaction rate.

The enthalpy change for such processes is $\Delta H = \Delta G + T\Delta S$. The term ΔG is necessarily negative, if the adsorption does in fact occur; but what can be said of ΔS? In all cases of *physical adsorption,* where only weak molecular forces are involved, the process is similar to simple condensation; the molecules have less freedom on the surface, and ΔS is negative. In the case of chemical adsorption, chemical bonds are involved, and an adsorbed molecule can become dissociated. In all but a very small number of cases, such dissociation is insufficient to render the overall entropy change positive. (The adsorption of hydrogen molecules onto a sulfur-contaminated iron surface is one of the few examples where ΔS is positive.) We can now state that all adsorptions with negative entropy change, which comprise all physical and the great majority of chemical adsorptions, are exothermic. This conclusion follows directly from Equation 6.3.

6.4.2 Evaporation Phenomena

When a liquid boils, the vapor pressure is equal to the atmospheric pressure. At any moment when the liquid is at the boiling temperature, vapor at 1 atmosphere pressure means that $\Delta H - T\Delta S = 0$, and so we may calculate the entropy change involved. At the boiling point, T_B, atmospheric pressure is in equilibrium with the liquid. Boiling liquid nitrogen may be considered as an example:

$$N_2(\text{liquid, 1 atm}) \underset{}{\overset{77.3K}{\rightleftharpoons}} N_2(\text{gas, 1 atm}).$$

Provided there is no superheating, the process is reversible and $\Delta G = 0$. This means that $\Delta G = \Delta H - T\Delta S = 0$, and so we may calculate the entropy change involved. At the boiling point T_B:

$$\Delta S = \frac{\Delta_{vap}H}{T_B} = \frac{5690 \text{ J mol}^{-1}}{77.3 \text{ K}}$$

$$= 73.6 \text{ J mol}^{-1} \text{ K}^{-1}.$$

If any further heat is added to the system, because the Dewar flask is imperfect, the liquid temperature will rise very slightly above the boiling point. As soon as this occurs, $\Delta G = \Delta H - T\Delta S$ becomes slightly negative, and the system redresses the imbalance by evaporating further. This then cools the liquid to the equilibrium temperature.

Entropies of vaporization at the boiling points and at atmospheric pressure have values near 88 J mol^{-1} K^{-1} (Table 6.1). This was investigated empirically by Trouton, and the rule was named after him. In most cases, the entropy increase is due only to the isothermal randomization of the molecules from the liquid to the vapor at 1 atm. In the case of water, because there is significant hydrogen bonding in the liquid, the entropy of liquid water is smaller than it would otherwise be. As a result, the relative increase in elbow room that the water molecules feel on evaporation is greater, and this is reflected in the high value of 109.5 J mol^{-1} K^{-1} for the entropy change.

Many millions of gallons of water are lost annually from lakes and rivers the world over from spontaneous evaporation. In regions in Australia where water conservation and control are important, monomolecular layers of fatty acids and similar film-forming materials are used to limit evaporation. Why, and to what extent, is this evaporation, which takes place well below the boiling point, spontaneous? First, let us assume that a dry air is blowing over the lake. At a temperature of 30°C, the water has a vapor pressure of 31 mm Hg. The latent heat of evaporation is roughly

TABLE 6.1
Some Entropies of Vaporization

Substance	Boiling Point (K)	$\Delta_{vap}H$ (kJ mol^{-1})	$\Delta_{vap}S$ (J mol^{-1} K^{-1})
N_2	77.3	5.69	73.6
$(C_2H_5)_2O$	308	27.2	88.3
C_6H_6	353	31.4	88.7
$C_6H_5 \cdot CH_3$	384	34.0	87.8
Hg	630	57.0	84.7
H_2O	373	40.9	109.5

+40.9 kJ mol^{-1}, which corresponds to a highly endothermic process. At the boiling point of water, the entropy term $T\Delta S = 373\ (109.5)/1000 = 40.9$ kJ mol^{-1} would be just sufficient to make ΔG marginally negative; but the temperature of the Australian outback is a mere 30°C (303 K).

The situation could be expressed like this:

$$\Delta G = \Delta H - T\Delta S$$

(probably small, but negative)

$$+40.9 \quad (303)(?).$$

This provides a value for ΔS near $40{,}900/303 = 135$ J mol^{-1} K^{-1}. This means that, for the process to occur, the entropy change for the process:

$$H_2O(\text{liq., 1 atm, 30°C}) \rightarrow H_2O(\text{dilute vapor in air, 1 atm, 30°C})$$

must be larger than 135 J mol^{-1} K^{-1}. This value, significantly larger than the value for $\Delta_{vap}S$ quoted above, can easily be achieved, but only by producing a sufficiently dilute vapor. Because spontaneous evaporation into a dry atmosphere causes cooling of the remaining water, the damp clothes on the washing line freeze in cool weather, and early peoples were afforded the equivalent of modern refrigeration. The early Egyptians, living in the exceptionally dry desert air, were probably quite familiar with ice manufactured by this method.

6.4.3 ENDOTHERMIC CHEMICAL PROCESSES

The previous examples have been mainly physical processes, but chemical ones are equally amenable to analysis in terms of free energy. The reaction that comprises dehydration by thionyl chloride was mentioned previously. It is endothermic, but proceeds with great certainty nonetheless. It would seem probable that ΔG is fairly large, in addition to having a negative sign. The reaction may be represented as:

$$SOCl_2(l) + H_2O(l) \rightarrow SO_2(g) + 2HCl(g).$$

The perhaps surprising spontaneity of this reaction is closely linked to the fact that copious gaseous products are given off by purely liquid reactants. The entropy will be seen to increase markedly. For this process, ΔS^0_{298} is easily determined from values of ΔS^0_{298} obtained from Appendix III:

$$\Delta S^0_{298} = \sum_{\text{products}} S^0_{298} - \sum_{\text{reactants}} S^0_{298}$$

$$= S^0(SO_2) + 2S^0(HCl) - S^0(SOCl_2) - S^0(H_2O)$$

$$= 248.2 + 2(186.9) - 215.7 - 69.9 = +336.4 \text{ J mol}^{-1} \text{ K}^{-1}.$$

The value for ΔH^0_{298} for the same reaction is calculated from the standard enthalpies of formation of the various chemicals, also to be found in Appendix III:

$$\Delta H^0 = \Delta_f H^0(SO_2) + 2\Delta_f H^0(HCl) - \Delta_f H^0(SOCl_2) - \Delta_f H^0(H_2O)$$

$$= +50.0 \text{ kJ mol}^{-1}.$$

Then, at 298 K:

$$\Delta G^0 = \Delta H^0 - T\Delta S^0$$

$$= +50.0 - 298(+336.4) / 1000$$

$$= -50.2 \text{ kJ mol}^{-1} \text{ SOCl}_2.$$

It is important to realize the full meaning of ΔG^0, consequent upon our using standard state data for entropies (S^0) and enthalpy changes (ΔH^0). In both cases, the superscript ^{0} infers that all materials, whether reactants or products are present in their standard states; that is, in effect, the stable states in which one handles them in laboratories at 1 bar pressure. The full significance of the result may be summarized by saying that, for the reaction in which 1 mol of liquid water is in contact and reacts with liquid thionyl chloride to produce sulfur dioxide and hydrogen chloride, each present as a gas at 1 bar partial pressure, the Gibbs free energy change at 298 K is −50.2 kJ.

The link between Gibbs free energy and the equilibrium constant can be appreciated qualitatively at this stage. The equilibrium constant in terms of pressures is given by:

$$K_P = p_{SO_2} p_{HCl}^2.$$

The partial pressures of the liquid reactants are invariant, and do not appear explicitly. It is apparent that a high applied pressure will discourage the reaction from proceeding further, and that a low pressure will enhance it. At a sufficiently high pressure, the system could be brought to equilibrium, and the process made reversible. Consequently, K_P is greater than 1. It is intuitively feasible that a negative value of ΔG^0 is associated with a value of K, greater than unity. Conversely, positive values of ΔG^0 would imply small values of K, usually less than 1. The relationship

$$\Delta G^0 = -a \log K$$

would fulfill these simple requirements, where a is a constant term. In the next chapter, such a relationship will be derived exactly.

As a second example of an endothermic, but spontaneous, process, we shall discuss the decomposition of cupric oxide to cuprous oxide and oxygen. The equation for the reaction and relevant data are:

$$2CuO(s) \rightleftharpoons Cu_2O(s) + \tfrac{1}{2}O_2(g)$$

$$\Delta H^0_{298} = +138.1 \text{ kJ mol}^{-1}$$

$$S^0_{298}(CuO) = 42.6 \text{ J mol}^{-1} \text{ K}^{-1}$$

$$S^0_{298}(Cu_2O) = 92.6 \text{ J mol}^{-1} \text{ K}^{-1}$$

$$S^0_{298}(O_2) = 205.1 \text{ J mol}^{-1} \text{ K}^{-1}.$$

From this we find that $\Delta S^0 = 110.0$ J mol^{-1} K^{-1}. This is, therefore, a reaction torn by internal conflict; the entropy change is favorable, but the energy change is strongly endothermic, and therefore unfavorable.

The actual stability of cupric oxide depends on the temperature. At 298 K, $T\Delta S^0$ is equal to 298 (+110.0)/1000 = 32.80 kJ mol^{-1}. Thus, at room temperature,

$$\Delta G^0 = +138.1 - 32.80 = +105.3 \text{ kJ mol}^{-1}.$$

Decomposition does not tend to occur, and cupric oxide is stable. However, at approximately 1256 K, the temperature is high enough to cause ΔG^0 to change sign, and at 1300 K, $T\Delta S^0$ has the value of 143.0 kJ mol^{-1}, with the result that ΔG^0 is –4.9 kJ mol^{-1}, and decomposition is favored.

6.4.4 Exothermic Chemical Process

In Germany in 1899, Ludwig Mond developed a process for extracting and purifying nickel. In the so-called "Mond process," raw impure nickel is first treated with CO gas, at atmospheric pressure and a temperature of 60°C to evaporate nickel as a carbonyl gas:

$$Ni(s) + 4CO(g) \underset{}{\overset{60°C}{\rightleftharpoons}} Ni(CO)_4(g).$$

Impure

From the data in Appendix III, we find that $\Delta H^0_{298} = -159.4$ kJ mol^{-1} and $\Delta S^0_{298} = -404.9$ J mol^{-1} K^{-1}. (The student should check this.)

The enthalpy change is favorable, but the entropy change is unfavorable for this reaction.

At 60°C, $\Delta G^0_{333} = \Delta H^0 - T. \Delta S^0 = -159.4 - 333(-404.9)/1000 = -24.6$ kJ mol^{-1}, which means that the enthalpy changes determine the outcome of this reaction. However, at higher temperatures we might expect that the entropy change decides.

In the second stage, the temperature of the gas, therefore, is increased to 200°C to decompose the nickel carbonyl gas into pure metallic nickel and CO:

$$Ni(CO)_4(g) \underset{}{\overset{200°C}{\rightleftharpoons}} Ni(s) + 4CO(g)$$

Pure

For the reversed reaction in the second stage, we find that:

$$\Delta G^0_{473} = \Delta H^0 - T\Delta S^0 = +159.4 - 473(+404.9)/1000 = -32.1 \text{ kJ mol}^{-1},$$

which means that $Ni(CO)_4(g)$ is unstable at higher temperatures.

To summarize, we may say that the sign of the Gibbs free energy change, which determines the direction of a process, depends on the two terms, ΔH^0 and $T\Delta S^0$, appearing in Equation 6.3. In many cases, they reinforce one another, but occasionally they tend to nullify each other. A change of temperature might then alter the sign of ΔG^0 and, therefore, the feasibility of the process.

6.5 STANDARD FREE ENERGY CHANGES

Values of ΔG^0 are required for a multitude of reactions at many different temperatures and, in general, one can calculate these from the appropriate values of ΔH^0 and S^0. Today, such calculations can be carried out very accurately with thermodynamic computer programs in a minimum of time. For example, let us have a closer look at the decomposition of cupric oxide (Section 6.4.3). If we take into account the temperature dependence of ΔS^0 and ΔH^0, the outcome of a computed calculation with the program HSC Chemistry [1] gives a temperature of 1393 K, above which ΔG^0 becomes negative. This temperature is, of course, more accurate than that given in Section 6.4.3, and the difference is quite considerable.

It would be unnecessarily tedious to tabulate data for all possible reactions; ΔG^0 values for formation reactions at 298 K are nonetheless very useful. These may be added or subtracted (ΔG is a function of state) to provide data for the reaction of interest, and corrections then made if the reaction temperature is higher (or lower) than the standard temperature of 298 K. (Such methods are elaborated in Chapter 10.) Appendix III includes a basic selection of such data; values for elements in their standard states are, of course, zero. Values for most, but not all, compounds are negative. The fact that values are positive in some cases does not imply that the compound (e.g., diborane, B_2H_6) cannot be made. It does show, however, that the compound is "thermodynamically unstable," and will decompose if time or catalysts are available.

REFERENCE

1. HSC Chemistry–Chemical Reaction and Equilibrium Software. Outotec, Finland. www .outotec.com/hsc.

PROBLEMS

6.1   Two moles of ideal gas expand irreversibly from 4 to 20 L at 54°C. Calculate ΔS for this change and, hence, ΔG.

6.2   Taking into account Trouton's rule: $\Delta S \approx 88$ J K^{-1}, estimate the normal boiling points of the following materials, the vaporization enthalpies for which are known:

	SnCl$_4$	cyclo-C$_6$H$_{12}$	(CH$_3$)$_2$CO	C$_2$H$_6$
$\Delta_{vap}H$ (kJ mol^{-1})	33.05	30.06	29.08	15.94

6.3   In a quartz-halogen lamp, the tungsten, which evaporates from the filament, is returned to the filament by a chemical transport mechanism. We can picture this as follows: The evaporated tungsten, from the hot filament, diffuses with an inert gas (argon) in the direction of the envelope. In colder parts of the lamp, gaseous tungsten reacts with a halogen or halogen compound to form a volatile compound. The reaction products diffuse back to the filament. In certain temperature zones, the volatile compound decomposes partly or completely, depending on its stability, during which tungsten is liberated and deposits on the filament. The halogen or halogen compound can then do its job again. In this way, no tungsten deposits on the quartz envelope and the total mass of the filament does not change. Nowadays, small amounts of dibromomethane are added to a quartz-bromine lamp. These decompose into carbon and hydrogen bromide. The carbon is taken up by the tungsten and hydrogen bromide reacts with gaseous tungsten:

$$W(g) + 5HBr(g) \rightarrow WBr_5(g) + \tfrac{5}{2}H_2(g).$$

The WBr$_5$ vapor then decomposes and deposits tungsten on the filament. Find:

(a) ΔG^0 for the above equation at 700 K, the normal temperature of the quartz envelope.

(b)   The minimum temperature of the filament at which WBr$_5$(g) will decompose into W(s) and Br$_2$(g).

6.4   If glass were used in place of quartz in a quartz-iodine lamp (see previous question), would there be any danger of the following reaction occurring at 350°C?

$$Na_2O + I_2(g) \rightarrow \tfrac{1}{2}O_2(g) + 2NaI(s)$$

(in glass)

Assume ΔH^0 and ΔS^0 are temperature-invariant. (In fact, quartz is widely used. What other factors have led to this choice?)

6.5 Palladium oxide, PdO, is moderately stable. Over the temperature range 298–1200 K, the standard free energy of formation in $J\ mol^{-1}$ is given by the approximate equation:

$$\Delta_f G^0(PdO(s)) = -108{,}046 + 93.36T.$$

What are ΔH^0 and ΔS^0 from this equation? Does PdO become more or less stable as the temperature increases? Investigate the situation at 1100 and 1200 K for the decomposition of PdO.

6.6 Chlorine reacts with methane at laboratory temperatures:

$$CH_4(g) + Cl_2(g) \rightleftarrows CH_3Cl(g) + HCl(g).$$

By determining ΔG^0_{1000} at 1000 K from the following free energies of formation, find out if this reaction will occur at this temperature.

Substance	$CH_4(g)$	$CH_3Cl(g)$	$HCl(g)$
$\Delta_f G^0_{1000}$ (kJ)	+19.445	+8.193	−100.79

6.7 The Scientific Group Thermodata Europe Substance database includes the following data on the reaction:

$$Si(s) + SiCl_4(g) \rightleftarrows 2SiCl_2(g):$$

$$\Delta H^0_{1500} = +297.2\ kJ\ mol^{-1}, \ \Delta G^0_{1500} = -19.305\ kJ\ mol^{-1}.$$

(a) What is the ΔS^0 at 1500 K?
(b) Is silicon stable to $SiCl_4$ at lower temperatures?
(c) The value of ΔH^0 is perhaps higher than expected. Is it reasonable in the light of bond energy arguments?

6.8 For the dissociation reaction:

$$PCl_5(g) \rightleftarrows PCl_3(g) + Cl_2(g)$$

at 472 K, $\Delta G^0 = +7.783\ kJ\ mol^{-1}$, but at 524 K, it has changed to $-0.682\ kJ\ mol^{-1}$. What qualitative changes does this indicate? Does Equation 6.3 lead you to expect such a change? Make what checks you can on these data.

7 Chemical Equilibrium

This chapter is concerned with the change in ΔG during the course of a reaction. During such a change, materials are constantly reacting and re-forming, but the trend will always be in the direction for which ΔG is negative. Finally, a certain combination of active masses will be reached where ΔG will be zero, and equilibrium achieved. This state is characterized by the equilibrium constant, which will have a fixed value at a given temperature. Before these calculations may be carried out in the simplest manner, the idea of activity, which embraces the concepts of gas pressure, solution concentration, and active masses of pure solids and liquids, must be cast in its unifying role. Also, the variation in free energy with change of activity, or "active mass," will be calculated.

7.1 PREAMBLE

We have seen that a reaction will proceed just as long as ΔG is negative, although the reaction may take place extremely slowly. Furthermore, a sufficient criterion for the state of equilibrium, where no net reaction in either direction occurs, is that $\Delta G = 0$.

It is well known that the reaction involving the hydrolysis of ethyl acetate to ethyl alcohol and acetic acid is a so-called "equilibrium reaction," in the sense that measurable quantities of all four components may be detected in the reaction mixture after an indefinite time. What is not so widely known, perhaps, is that *all* reactions are equilibrium reactions, and that no reaction goes literally to completion. Even the formation of liquid water from gaseous hydrogen and oxygen is not complete, and there is a certain finite pressure of hydrogen in equilibrium with the liquid. The fact that in the air over the Atlantic Ocean only five hydrogen molecules, on average, are in equilibrium with it does not detract from the main point that *no reaction* is complete. Consider in more detail the hydrogenation of ethylene; the equation is:

$$C_2H_4(g) + H_2(g) \xrightleftharpoons{\text{catalyst}} C_2H_6(g).$$

$$\text{1 bar} \quad \text{1 bar}$$

Suppose the initial partial pressures to be 1 bar for each reactant. (The partial pressure of one component of a mixture of gases is defined as the pressure it would exert if it were alone in the available space. For an ideal mixture of perfect gases, the total pressure is the sum of the various partial pressures.) At 25°C, ΔG^0 for this reaction is −100.4 kJ mol^{-1}, and so the driving force is considerable, and reaction takes place over the catalyst with great vigor. The temperature of the reaction vessel is kept constant. As the reaction occurs, the partial pressures of ethylene and hydrogen

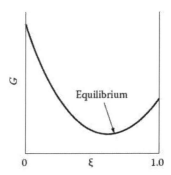

FIGURE 7.1 The Gibbs free energy, G, as a function of the extent of the reaction ξ.

drop, and their reduced active mass, or presence, causes the value of ΔG to become a smaller negative number. This situation is aggravated by the increasing active mass of the product, ethane, and the associated back reaction. Eventually, the active mass of the reactants is so low, and that of ethane so high, that the forward rate exactly equals the rate of the backward reaction and no net reaction occurs, so $\Delta G = 0$.

Let $dn(C_2H_4)$, $dn(H_2)$, and $dn(C_2H_6)$ be the changes in the amount of moles of C_2H_4, H_2, and C_2H_6. A useful variable is the extent of the reaction ξ (Figure 7.1). In this case, we may write:

$$dn(C_2H_4) = dn(H_2) = -dn(C_2H_6) = d\xi.$$

If $\xi = 0$, no reaction occurs so there are only reactants. If $\xi = 1$, at least one of the reactants has completely reacted. A large negative ΔG^0 gives a minimum on the right-hand side of the curve, as is the case in the above reaction. A large positive ΔG^0, however, gives a minimum on the left-hand side, and leads to a very small equilibrium constant. It seems plausible that a large negative value for ΔG^0 should lead to near complete reaction from left to right, and also to a large equilibrium constant. This relationship we shall derive.

Before we can measure the changes in ΔG during the course of a reaction, we must find the link between G and pressure, concentration, and physical state.

7.2 VARIATION OF G WITH GAS PRESSURE

We start with the definition of G:

$$G = H - TS.$$

By substituting

$$H = U + PV,$$

we have

$$G = U + PV - TS.$$

Complete differentiation of this yields:

$$dG = dU + PdV + VdP - TdS - SdT.$$

However, as

$$dU = dq - PdV,$$

and

$$dS = dq/T$$

for a reversible change, we may substitute

$$dU = TdS - PdV.$$

This immediately looks much more promising, as:

$$dG = (TdS - PdV) + PdV + VdP - TdS - SdT$$

or

$$dG = VdP - SdT \tag{7.1}$$

This simple equation has great subsequent application and should be memorized. For our particular need, we are interested in the effect of pressure, and can put $dT = 0$ (constant temperature), so:

$$\left(\frac{dG}{dP} \right)_T = V.$$

Before going on to integrate this relation, it is rewarding to introduce a new term, the chemical potential, which is closely related to free energy.

7.2.1 Chemical Potential, μ

For a pure substance or a system where the chemical composition is constant, Equation 7.1 holds. If material is added to or taken from the system or, more important, materials are consumed or produced by a chemical reaction, then we are unable completely to specify the system at every stage of the alteration. The thermodynamicist reacts to this situation much as nature to a vacuum; the gap must be filled. In addition to temperature and pressure, which have sufficed until now to describe the system, we must now introduce a new variable into our repertoire. The amount of each material present will now figure prominently. *Chemical potential* is defined here as:

$$\mu_i = \left(\frac{\partial G}{\partial n_i}\right)_{T,P,n_j\ldots}.$$

(7.2)

That is, the chemical potential of the component, i, is the partial derivative of G, with respect to the number of moles of i, written as n_i; temperature, pressure, and the amounts of all other components, n_j, etc., are held constant. It follows from this that where before Equation 7.1 we wrote $dG = VdP-SdT$, changes in G could now arise due to changes in n_1, n_2 the number of moles of components 1, 2, etc., and we have to write:

$$dG = VdP - SdT + \left(\frac{\partial G}{\partial n_1}\right)dn_1 + \left(\frac{\partial G}{\partial n_2}\right)dn_2 + \cdots.$$

(7.3)

Here, the additional terms account for new material. Summation at constant temperature and pressure gives simply:

$$G = \sum_{\text{all } i} \mu_i n_i.$$

(7.4)

The rather cumbrous definition of μ given in Equation 7.2 will be more acceptable when it is realized that for a pure material the free energy varies linearly with the number of moles (Figure 7.2), and that:

$$\mu = \frac{G}{n}.$$

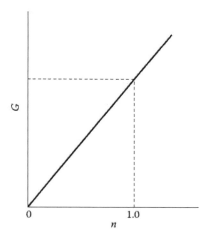

FIGURE 7.2 The Gibbs free energy, G, for a pure component as a function of the total number of moles, n.

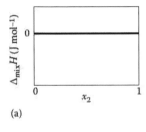

(a)

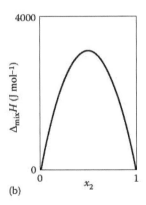

(b)

FIGURE 7.3 $\Delta_{mix}H$ as a function of the mole fraction (x_2) of component 2 for (a) an ideal and (b) a nonideal solution.

In this case, μ is merely the molar free energy, or free energy per mole, written as G_m. (To write G for a compound does not imply measurement in absolute terms. The value is the Gibbs free energy related to the elements, which are arbitrarily assigned zero values. The term G is thus identified with $\Delta_f G^0$ for the compound.) This simple equation is all that is needed to define μ, provided that the mixture behaves ideally. Unfortunately, no mixture is ideal in the sense that each molecule sees all surrounding molecules as identical neighbors. There are interactions between them, and usually A–A interactions are unlike A–B interactions, and usually these are both unlike those between B and B. In the few cases of near-ideal mixtures, we find great chemical similarities, such as for the benzene/toluene, and ethylene bromide/propylene bromide systems. For these mixtures, ΔH and ΔV mixing are small but nonzero (Figure 7.3). (For truly ideal solutions, ΔH and ΔV are zero.) The magnitude of these quantities is therefore a measure of the nonideality of the mixture.

Table 7.1 shows the wide variation that exists in solution behavior. The acetone/carbon disulfide system is far from ideal; this nonideality will also show up in properties such as the vapor pressure curves, and the boiling point diagram. As a result

TABLE 7.1
ΔH of Mixing 1:1 Ratios, to Form 1 mol of Mixture

Mixture	Temperature (K)	$\Delta_{mix}H$ (J mol^{-1})	Reference
Benzene/toluene	353	46	a
Benzene/CCl$_4$	298	109	b
Acetone/CS$_2$	308	1460	b

[a] *Source:* Cheeseman, C.H., and Ladener, W.R., *Proc. R. Soc.*, A229, 387, 1955. With permission.
[b] *Source:* Staveley et al., *Trans. Faraday Soc.*, 51, 323, 1955. With permission.

of all this, we could say that μ_i for a compound i (the subscript implies a mixture) depends not only on the nature of the compound, *but also on the environment in which it finds itself.* In this more complete sense, μ_i is known as the *partial molar free energy,* and may be pictured as the change of free energy upon adding 1 mol of the material to a very large and invariant system.

Consider 2 mol of pure acetone, and suppose that the Gibbs free energy is G_{acet}. μ is simply $G/n = G_{acet}/2$, and the free energy of the acetone sample is 2μ. Alternatively, consider a lake as the system to which 2 mol acetone is added. By how much does the free energy of the system increase?

Temperature and pressure may be supposed constant and, for the water molecules at least, their environment is unchanged. Acetone, however, now finds itself in a pure water environment, and μ_{acet} will be (Figure 7.4):

$$\left(\frac{\partial G}{\partial n_{acet}} \right)_{T,P,n_{H_2O}},$$

and this may well be different from G/n.

Using Equation 7.3, which says in this case that:

$$dG = \left(\frac{\partial G}{\partial n_1} \right)_{T,P,n_2} dn_1,$$

we have

$$dG = 2 \left(\frac{\partial G}{\partial n_{acet}} \right)$$

$$= 2\mu_{acet}.$$

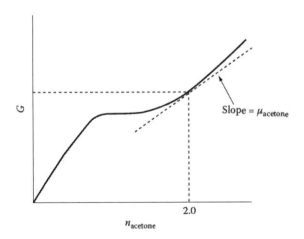

FIGURE 7.4   A hypothetical Gibbs free energy curve as a function of the number of moles of acetone added to water.

(Again, we emphasize that the subscript to μ implies a mixture.)

Now, let us turn to a system of two components miscible in all proportions. The components could be gases, liquids, or solids. The chemical potential of a component can conveniently be displayed using a graphical representation [1]. The molar free energy of the solution G is defined as:

$$G_m = \frac{G}{n_A + n_B},$$

where n_A is the number of moles of A, and n_B is the number of moles of B. Differentiation of G with respect to n_A yields the chemical potential μ_A:

$$\mu_A = G_m + (n_A + n_B) \cdot \left(\frac{\partial G_m}{\partial n_A}\right)_{n_B}.$$

After some manipulation and differentiation, the following expression can be derived:

$$G_m = \mu_A + x_B \frac{dG_m}{dx_B}, \qquad (7.5)$$

where x_B is the mole fraction of component B in the system. From this expression, the chemical potential of each component in a two-component system is determined as is shown in Figure 7.5.

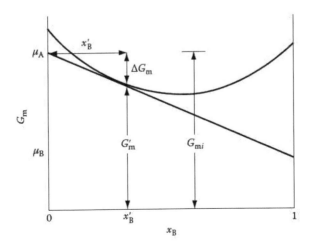

FIGURE 7.5 The Gibbs free molar energy of a two-component solution as a function of composition at a fixed temperature.

For a given composition, x'_B, the corresponding Gibbs molar free energy of the solution is G'_m. The tangent slope at (x'_B, G'_m) is $-\Delta G_m /x'_B$. The Gibbs molar free energy of solution at the intercept corresponding to the A component is $G_{m,I} = G'_m + \Delta G_m$. When Equation 7.5 is evaluated at x'_B, the intercept at $x_B = 0$ equals μ_A in the solution. Similarly, if Equation 7.5 is evaluated for the B component, the intercept at $x_B = 1$ equals μ_B. So, the slope of G versus x_B is equal to the difference in the chemical potential of the two components in solution $(\mu_B - \mu_A)$. This will become important when phase separation is considered in partially miscible liquids.

7.2.2 Pressure and Chemical Potential for Ideal Gases

At the end of Section 7.2, we concluded that:

$$\left(\frac{dG}{dP}\right)_T = V, \quad \text{or} \quad dG = VdP.$$

Using molar quantities, we can instead write:

$$d\mu = V_m dP,$$

where V_m is the volume of 1 mol. Now, assume that the equation of state holds, and $V_m = RT/P$. Then,

$$d\mu = RT\frac{dP}{P} \quad (T \text{ constant}) \tag{7.6}$$

On integrating this between two states, 1 and 2, we have:

$$\mu_2 - \mu_1 = RT(\ln P_2 - \ln P_1)$$

Or $\qquad\qquad\qquad\qquad\qquad\qquad (T \text{ constant}) \qquad\qquad\qquad (7.7)$

$$\mu_2 = \mu_1 + RT \ln P_2/P_1.$$

Thus, 1 could represent some reference state, with a chemical potential of μ_1. Let us denote this reference potential and pressure by θ. This means that we can dispense with 1 and 2, and write:

$$\mu = \mu^\theta + RT \ln P/P^\theta \quad (T \text{ constant}) \tag{7.8}$$

The choice of this reference state or "starting point" is, of course, quite arbitrary, but the criterion of ease of working is usually adopted. The most generally useful reference chemical potential is that at a pressure of 1 bar. This will be denoted by a superscript zero, so we now have

TABLE 7.2

Changes of Chemical Potential Resulting from Nonunit Pressures Expressed as $(\mu - \mu^0)$, for an Ideal Gas at 298 K

Pressure (bar)	Change of Chemical Potential, $\mu - \mu^0$ (kJ)
0.01	−11.42
0.05	−7.40
0.1	−5.69
0.5	−1.72
1	0
2	+1.72
10	+5.69
20	+7.40
100	+11.42

$$\mu = \mu^0 + RT \ln (P(\text{bar})/1 \text{ bar}) \quad (T \text{ constant}). \tag{7.9}$$

The physical reality of this equation is seen best in terms of a table of $(\mu - \mu^0)$ values at various pressures (Table 7.2).

Before moving on to real, and therefore nonideal, gases, let us take stock. Equation 7.9 shows us that the chemical potential of a pure gas varies with pressure, that μ^0 is a reference value, and that the logarithmic term allows for nonunit pressure. A dimensional check reveals that the logarithmic term has no units, and that μ has the units of RT, that is, joules per mole. So far, so good.

7.2.3 CHEMICAL POTENTIAL FOR REAL GASES

Of course, no gas is truly ideal; real gases follow the $PV = RT$ relationship to a greater or lesser extent, but are unable to follow it exactly because of their physical constitution. Molecules interact with each other and they have a volume of their own. Nonetheless, real gases come very near to ideality at low pressures, and it can be said that the equation of state is a "limiting law," that is, for 1 mol:

$$\lim_{P \to 0}(PV) = RT.$$

Many industrial processes occur only at high pressures, and so we must come to grips with this problem of nonideality.

In formulating Equation 7.6, the assumption of ideality was made (unjustly as it turns out) and V_m put equal to RT/P. Therefore, Equations 7.8 and 7.9, which follow from it, also have the assumption of perfection within them, and they, too, will therefore fail to describe a real gas. There are now two courses of action that we could take. First, we could substitute for V_m a more complicated expression derived from the van der Waals equation, or another empirical equation of state. This would be a lengthy undertaking and as new data become available, fresh integrations would have to be performed. The second alternative would be to accept Equation 7.8 on one condition: that in place of P, we put a new quantity *that revalidates this relationship.* Ideally, it would not vary much from the ideal pressure and could be approximated as such when the pressure was low enough. The second alternative is adopted because, on the whole, it is less onerous. The new function is called *fugacity,* which is derived from the Latin *fugere,* to flee, and means literally "escaping tendency." It is denoted by f. Using this definition, for a real gas, we can rewrite Equation 7.8 as:

$$\mu = \mu^{\ominus} + RT \ln \frac{f}{f^{\ominus}} \qquad (T \text{ constant}). \qquad (7.10)$$

In many cases, but not all, $f^{\ominus}$ is assigned a value of 1 bar. (Note that the reference fugacity is 1 bar, and that the pressure may well differ from this value.)

We now have a standard state for which a zero superscript is applicable:

$$\mu = \mu^0 + RT \ln \frac{f \text{ (bar)}}{1 \text{ (bar)}} \text{ and } f = \exp\left(\frac{\mu - \mu^0}{RT}\right),$$

where f is to be considered an effective pressure.

A formal definition would be double-headed. First, Equation 7.10 is a true expression of the chemical potential of a real gas and, second, fugacity and pressure become identical at zero pressure (Figure 7.6):

$$\lim_{P \to 0}\left(\frac{f}{P} \to 1\right).$$

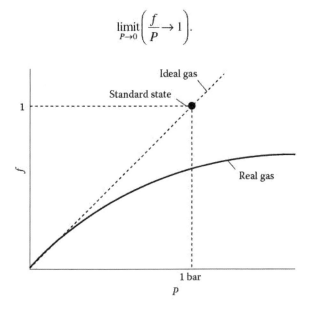

FIGURE 7.6 Definition of standard state of a real gas with dimension of fugacity.

Fugacity is not, repeat not, the actual pressure. It must not only account for the actual behavior of real gases, but it must also overcome the erroneous assumptions of perfection that are still part of the fabric of the equation. It, therefore, overcorrects.

From the definition of the chemical potential of 1 mol pure gas

$$\mu = \mu^0 + RT \ln f,$$

it can be derived that:

$$\ln \frac{f}{P} = \int_0^P \left(\frac{V}{RT} - \frac{1}{P} \right) dT.$$

With this equation, f can be calculated at a particular temperature and pressure. This can be done if the molar volume V of the gas is known as a function of the pressure P at the specified temperature. To give an example, nitrogen at 25°C and 1000 atm is well away from ideality. The ideal pressure (nRT/V) is 498 atm, but the fugacity is, in fact, 1839 atm.

At low pressures, real gases come very near to ideality and so by approximation we obtain:

$$\frac{f}{P} \approx \frac{P}{P_{ideal}},$$

where P_{ideal} is the pressure an ideal gas with the same molar volume as the real gas would have.

Example 7.1

At 10 atm and 0°C, nitrogen has a molar volume of 2.232 L. Estimate the fugacity of nitrogen in these circumstances.

First, we must calculate P_{ideal}:

$$P_{ideal} = \frac{RT}{V} = \frac{(8.314)(273.15)}{2.232 \times 10^{-3}} \frac{1}{1.013 \times 10^5} = 10.043 \text{ atm}$$

and so

$$f \approx \frac{P^2}{P_{ideal}} = \frac{10^2}{10.043} = 9.957 \text{ atm.}$$

7.2.4   ACTIVITY

The situation is streamlined somewhat by substituting an abbreviated form of the fugacity ratio $\dfrac{f}{f^{\theta}}$. We define the activity, a, as $a = \dfrac{f}{f^{\theta}}$. As a result, Equation 7.10 can be made to look much more attractive. That is,

$$\mu = \mu^{\theta} + RT \ln a.$$

This equation may be expressed in words. The chemical potential is equal to a reference value μ^{θ}, associated with a fugacity f^{θ}, plus the correction term RT, $RT \ln \dfrac{f}{f^{\theta}}$. The ratio $\dfrac{f}{f^{\theta}}$ is abbreviated a. If we take $f^{\theta} = 1$ bar, then:

$$a = f/f^{\theta} = f/1.$$

The denominator is stressed, as it shows that a *is dimensionless*. We have now defined a standard state for which a zero superscript is applicable, and can write a simplified equation, which is widely useful:

$$\mu = \mu^{0} + RT \ln a \quad (T \text{ constant}). \tag{7.11}$$

Here, the superscript zero implies, for real gases, a unit fugacity standard state. If, during an experiment, the fugacity happens to coincide with the standard state, then the activity is one, and the chemical potential is equal to μ^{0}.

So far, activity has been treated as a somewhat abstract parameter, but we can relate it to the actual pressure by means of an *activity coefficient* γ, when we state that:

$$a = \gamma P.$$

Obviously, γ is unity in the case of an ideal gas, and some other value for a real gas. Now, Equation 7.11 takes the form

$$\mu = \mu^{0} + RT \ln P + RT \ln \gamma.$$

Comparison of this with Equation 7.9 shows that the final term merely allows for nonideality. It is no more than a correction term. (When gaseous mixtures are involved, it is the various partial pressures p_i, or partial fugacities, f_i, which become important. In such a case, f_i would represent the fugacity of the compound, i, in the particular environment in which it found itself.)

The fugacity of an ideal gas is equal to its pressure, and all gases approach ideal gas behavior as the pressure is reduced. If the activity coefficient is greater than unity, the gas has a greater potential than if it were ideal at the same temperature and pressure. If the activity coefficient is less than unity, the gas has a lower activity and a lower chemical potential than if it were ideal.

7.3 THE ACTIVE MASS OF PURE LIQUIDS AND SOLIDS

When a pure solid or liquid takes part in a reaction, it does so to the extent that individual molecules are prepared to leave the mass of the material. This tendency is closely linked to the vapor pressure of the material, and is constant at a given temperature. The vapor pressure of a piece of nickel, say, although very small, is finite, and although this depends principally on temperature, there is a slight dependence on the *total* applied pressure. Generally we may write for a substance, i (pure or a component in a mixture in solid, liquid, or gas phase):

$$\mu_i = \mu_i^0 + RT \ln a_i,$$

where μ_i^0 is the chemical potential of a substance, i, in some standard state. The standard state of a pure liquid is the substance at pressure P^0. If the pressure deviates from P^0 and, assuming the substance to be incompressible, we may write Equation 7.6 in another form:

$$d\mu = \mu_i(P) - \mu_i(P^0) = \int_{P^0}^{P} V_i \, dP = V_i(P - P^0),$$

so that the activity at pressure P is given by:

$$RT \ln a_i = V_i(P - P^0) \text{ or } a_i = \exp\frac{V_i(P - P^0)}{RT}$$

(pure solid or liquid).

At any moderate pressure, the exponent is generally quite small, so that the activity of a pure solid or liquid can be taken as nearly equal to unity under ordinary conditions.

Example 7.2

Find the activity of pure liquid water at pressures of 3 bar and 1000 bar at 25°C. (The molar volume of water is 1.805×10^{-5} m³.) At 3 bar,

$$a_{H_2O(l)} = \exp\frac{(1.805 \times 10^{-5})((3-1) \times 10^5)}{(8.1314)(298.15)} = 1.0015.$$

At 1000 bar,

$$a_{H_2O(l)} = \exp\frac{(1.805 \times 10^{-5})((1000-1) \times 10^5)}{(8.1314)(298.15)} = 2.0894.$$

Thus, to summarize, for pure liquids and solids, μ is virtually constant, and the materials are assigned unit activity, that is $a = 1$, and $\mu = \mu^0$.

7.4 ACTIVITY OF MATERIALS IN SOLUTION

7.4.1 SOLVENTS

We have seen that the activity of a gas is related to its effective pressure. For liquids, it is the liquid vapor pressure that is important and, for solutions, the vapor pressure is linked to the concentration of the component of interest. The concentration of a solvent or major component is conveniently measured as the mole fraction, x. For ideal vapors and an ideal solution, it may be shown that the chemical potential of a component number 1 is:

$$\mu_1 = \mu^0_{(pure\ 1)} + RT \ln x_1 \quad \text{(ideal solution)}.$$

The term $\mu^0_{(pure\ 1)}$ represents the chemical potential of the pure material, where x_1 is unity. For nonideal solutions, a term a, the activity of the solvent, which may be thought of as the effective mole fraction, is introduced, and we have, for component 1:

$$\mu = \mu^0_{(a=1)} + RT \ln a.$$

The standard state of unit activity corresponds to pure solvent or unit mole fraction. (All solutions approach ideality at zero solute concentration.) Although the standard state is different, the form of this equation is identical to that of Equation 7.11. The activity may be linked with the actual mole fraction by the activity coefficient, γ; that is, $a = \gamma x$. For ideal solutions, $\gamma = 1$.

7.4.2 SOLUTES OR MINOR COMPONENTS

For solutes, molarity, m, defined as the number of moles of solute dissolved in 1000 g of solvent is a more convenient measure of concentration. For any given solvent it is approximately proportional to the mole fraction. An equation similar to Equation 7.11 may be derived for solutes:

$$\mu = \mu^0_{(a=1)} + RT \ln a,$$

where activity is a measure of the effective molality. For dilute solutions, the molality activity coefficient, γ, is unity and $a = \gamma m = m$. The standard state ($a = 1$) corresponds to the ideal state of unit molality. (Such an ideal state is hypothetical, because all solutions deviate from ideality. However, by extrapolating up from a very dilute, ideal region, the state is adequately described.)

7.5   A SUMMING UP: ACTIVITY AS A UNIFYING CONCEPT

(a) We have treated all manner of materials, solid, liquid, gas, pure and mixed.

(b) In each case, the *active mass* was identified with activity, *a*.

(c) A *standard state*—a starting point for each material—was assigned a value of *unit activity*.

(d) In each case, the chemical potential is given as:

$$\mu = \mu^0_{(activity=1)} + RT \ln a.$$

(e) The standard states, with $a = 1$ in each case, are summarized as follows:
Pure gas: $f = 1$ bar.
Pure liquids and solids: The pure materials have an activity of virtually unity over a wide range of pressures.
Gas mixtures: Ideal perfect mixture, $p_i = 1$ bar. Mixture of real gases, $f_i = 1$ bar.
Solid or liquid solutions: For the major component, $x = 1$. For the minor component, hypothetical ideal state, $m = 1$.

7.6   PRACTICAL ASPECTS OF ACTIVITY

Detailed tables of fugacity at different pressures are available for many gases. Luckily it is not necessary to make actual measurements for every gas (unless justified by a need for high accuracy). Most gases show similar behavior in the region of their critical points. (The critical points are the temperature and pressure at which the liquid–vapor coexistence curve in a P–T diagram of a pure substance terminates. Above the critical temperature, there can be no liquid phase.) These similarities are most easily seen in terms of activity coefficients, $\gamma = f/P$, if the pressures are expressed as *reduced pressures* (P/P_c, where P_c is critical pressure), and the temperatures as *reduced temperatures* (T/T_c, where T_c denotes critical temperature). As an example, it is found that the activity coefficient for methane at a reduced pressure of $P/P_c = 2$ and reduced temperature of $T/T_c = 2$ is near 0.94. This means that $f = 0.94P$. For trichlorosilane, under the corresponding conditions of $P/P_c = T/T_c = 2$, γ is also very close to 0.94. In this way, it is possible to draw up charts for all gases, showing fugacity ratio varying with reduced pressure, for some chosen value of the reduced temperature. Such a chart is shown as Figure 7.7. Resulting corrections are small, that is f/P is near 1, as long as the temperature is above twice the critical value, and the pressure below twice its critical value.

Example 7.3

Ethanol takes part in a reaction at 300°C, with a partial pressure of 30 atm. Table 10.1 gives the critical point data, which are $T_c = 516$ K, and $P_c = 63$ atm. What is the fugacity under these conditions? In this case, we have reduced temperature =

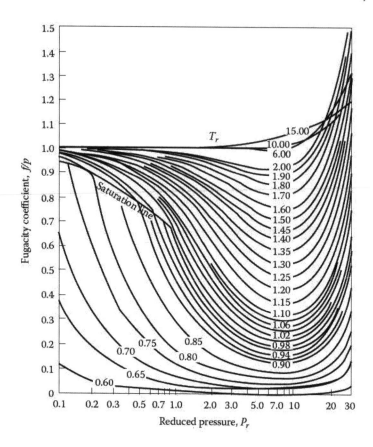

FIGURE 7.7   Generalized fugacity coefficient chart.

$T/T_c = 573/516 = 1.11$, and reduced pressure $= P/P_c = 30/63 = 0.48$. From Figure 7.7, we find $\gamma = f/P = 0.92$. Thus, $f = \gamma P = 27.6$ atm.

7.7   EQUILIBRIUM AND THE REACTION ISOTHERM

Appendix III includes values of ΔG^0 for a selection of formation reactions. When properly combined, these provide ΔG^0 data for a wide variety of other reactions. However, these values apply only to a hypothetical situation in which reactants and products are in their standard states, that is at unit activity. However, the free energy change can now be worked out for any other set of conditions, because we know how the chemical potential varies with activity. The vital relationship is, of course, $\mu = \mu^0 + RT \ln a$.

Consider a perfectly general reaction, such as:

$$aA + bB + \cdots \rightleftarrows pP + qQ + \cdots$$

in which aA represents "a moles of A," and so on. In a particular set of circumstances, we may be required to know whether the reaction will tend to occur. We may calculate ΔG^0 from $\Delta_f G^0$ data, but this refers only to standard states.

Let us start with what is known. Because ΔG^0 is G^0(products) minus G^0(reactants), we may use Equation 7.4 to write:

$$\Delta G^0 = \sum_{products} n_i \mu_i^0 - \sum_{reactants} n_i \mu_i^0$$

$$= p\mu_P^0 + q\mu_Q^0 - \left(a\mu_A^0 + b\mu_B^0\right)....$$

(7.12)

Similarly, for nonstandard states, we may write:

$$\Delta G = p\mu_P + q\mu_Q - (a\mu_A + b\mu_B)....$$

This equation may be expanded because Equation 7.11 links chemical potential to activities. Thus,

$$\Delta G = p\mu_P^0 + pRT \ln a_P$$

$$+ q\mu_Q^0 + qRT \ln a_Q ...$$

$$- (a\mu_A^0 + aRT \ln a_A)$$

$$- (b\mu_B^0 + bRT \ln a_B)....$$

(7.13)

The symmetrical arrangement helps us to recognize that the terms in the first column add up to ΔG^0 (see Equation 7.12). If now the numbers of moles, p, q, a, and b are put inside the logarithms, the equation may be rewritten as:

$$\Delta G = \Delta G^0 + RT \ln \frac{a_P^p a_Q^q ...}{a_A^a a_B^b ...}.$$

(7.14)

This is the *general reaction isotherm*, also known as the van't Hoff isotherm; it is of prime importance. The logarithmic ratio is sometimes known as the *activity quotient*, and is written as Q. As before, ΔG is a measure of the affinity of the process actually occurring, where the logarithmic term makes adjustment for nonunit activities. This equation would apply, for example, when it was required to determine the feasibility of a reaction for which all starting activities are known.

If now the reaction is allowed to proceed, the activities will alter, and finally the reaction will reach equilibrium. At this point, ΔG is zero, and so Equation 7.14 becomes:

$$\Delta G^0 = -RT \ln \frac{a_P^p a_Q^q ...}{a_A^a a_B^b ...}.$$

The activities now have an extra significance, because taken together they define the equilibrium condition, and are fixed among themselves. If T is fixed, then so is ΔG^0. This means that the activity term must also be constant; it is now equal to *the equilibrium constant* in terms of activities, K.

First critical comment. Because activity has been carefully defined as a dimensionless ratio, K must also be dimensionless. For gases, with the standard state defined as unit fugacity, $a = f/f^0 \sim p/p^0 = p$ bar/1 bar.

Finally, the standard equilibrium isotherm may be written down:

$$\Delta G^0 = -RT \ln K. \tag{7.15}$$

Despite its simplicity, this is probably the most widely useful equation in all of thermodynamics. It is now that we can appreciate the importance of ΔG^0 data; values for different reactions may be combined to give the equilibrium constant for any other reaction of interest.

Example 7.4

We shall now apply the concepts of activity and equilibrium that we have developed to a particular reaction. A laboratory study has provided information on the reaction:

$$FeO(l) + CO(g) \rightleftarrows Fe(s) + CO_2(g)$$

over the temperature range 1650 K (the melting point of FeO), to 1809 K (that of Fe). The data are expressed linearly in temperature:

$$\Delta G^0 = -40,524 + 37.87T \quad \text{in J mol}^{-1}.$$

This reaction includes crystalline, liquid, and gaseous phases, but even if solutions are present, the activities may be determined for each component. For simplicity, however, we shall assume that both condensed phases are pure and have unit activity.

By using these data, let us calculate the equilibrium constant at 1700 K. First, we find that ΔG^0_{1700} is +18.257 kJ mol^{-1}.

Second critical comment. The standard states are those appropriate to a temperature of 1700 K. These are, for carbon monoxide and carbon dioxide, the gases at unit fugacity, for iron the regular crystal stable at that temperature (alpha phase), and for ferrous oxide, the pure liquid.

Using Equation 7.15, we may write, for the equilibrium situation:

$$\Delta G^0 = -RT \ln \frac{a_{Fe} a_{CO_2}}{a_{FeO} a_{CO}}$$

$$18,257 = -8.314(1700) \ln \frac{1 \cdot p_{CO_2} (\text{bar})/1 \ (\text{bar})}{1 \cdot p_{CO} (\text{bar})/1 \ (\text{bar})}$$

Notice that because iron and ferrous oxide are assumed pure, their activities are put equal to one, and also that fugacity ratios for carbon monoxide and carbon dioxide are put equal to pressure ratios at this high temperature.

The pressure ratio is now seen to be identical to K_p, the equilibrium constant expressed in terms of relative pressures. (Note that, in some cases, the equilibrium constant K_c is applied to gaseous systems. This term is not needed at this stage, and leads to complications; extra terms involving $RT^{\Delta n}$, where Δn is the increase in the number of gaseous moles when the reaction goes from left to right, are brought in when expressing K_c in terms of K or K_p. The concept of partial pressures is both more simple and more rewarding, and will be used in preference in this book.)

Finally, we find that $K_p = p_{CO_2}/p_{CO} = 0.275$ so that carbon monoxide is only partly effective as a reducing agent at this temperature. So important is this kind of equilibrium that diagrams (named after Richardson) of equilibrium p_{CO_2}/p_{CO} ratios are plotted for many coke reduction processes over a wide range of temperature as an aid to process planning.

It is interesting that we can derive the same result from the Gibbs free energy G, as a function of the extent of the reaction ξ. Therefore, we are going to examine how the Gibbs free energy of this reaction mixture changes during the course of the reaction. Suppose we start with 1 mol of FeO(l) and 1 mol of CO(g). We can express the number of moles of each component during the course of the reaction in terms of the extent of reaction ξ. If $\xi = 0$, no reaction occurs, when the system consist of 1 mol FeO(l) and 1 mol CO(g). If $\xi = 1$, both reactants have completely reacted. The Gibbs free energy of the reaction mixture is given by

$$G_{system} = n_{FeO} \cdot \mu_{FeO} + n_{CO} \cdot \mu_{CO} + n_{Fe} \cdot \mu_{Fe} + n_{CO_2} \cdot \mu_{CO_2} \tag{7.16}$$

$$\mu = \mu^0 + RT \ln a.$$

For an ideal gas, $\mu = \mu^0 + RT \ln(p/1 \text{ bar})$ and for a pure liquid or solid, $\mu = \mu^0 + RT \ln (1) = \mu^0$.

$$G(\xi) = (1-\xi)\mu_{FeO}^0 + (1-\xi)\left\{\mu_{CO}^0 + RT \ln P_{CO}\right\} + \xi\mu_{Fe}^0$$
$$+ \xi\left\{\mu_{CO_2}^0 + RT \ln P_{CO_2}\right\} \tag{7.17}$$

If the reaction is carried out at a constant total pressure of 1 bar, then: $P_{CO} = x_{CO}P_{tot} = x_{CO}$ and $P_{CO_2} = x_{CO_2}P_{tot} = x_{CO_2}$. The total number of moles in the gas phase is $(1-\xi) + \xi = 1$ mol. Therefore, $P_{CO} = x_{CO} = (1-\xi)$ and $P_{CO_2} = x_{CO_2} = \xi$.

$$G(\xi) = (1-\xi)\mu_{FeO}^0 + (1-\xi)\left\{\mu_{CO}^0 + RT \ln (1-\xi)\right\} + \xi\mu_{Fe}^0$$
$$+ \xi\left\{\mu_{CO_2}^0 + RT \ln \xi\right\} \tag{7.18}$$

μ^0 values are derived from Equation 6.1

$$G = H - TS.$$

Absolute values of enthalpy of substances cannot be measured, but the enthalpy differences between two temperatures can be measured with a calorimeter.

From $C_p = (dH/dT)_{P,n}$ enthalpy can be calculated as:

$$H_T = H_{f,298.15} + \int_{298.15}^{T} C_p dT + \sum H_{trs} \tag{7.19}$$

where $H_{f,298.15}$ is the enthalpy of formation at 298.15 K, and H_{trs} is the enthalpy of transition of the substance. The enthalpy scale is fixed by defining $H = 0$ for the elements in their most stable state at 25°C and 1 bar, mainly because this is convenient for calculations at elevated temperatures. This is called the reference state. The enthalpy of compounds also contains their enthalpy of formation $\Delta_f H$ from elements. This is usually measured calorimetrically by letting pure constituent elements react and form compounds at 298.15 K and 1 bar. The enthalpy of the compound is therefore calculated by adding the enthalpy of formation to the experimental enthalpy difference $H_T - H_{298}$.

Absolute entropy values can be calculated from the experimental heat capacity values using Equation 7.20 and numerical integration:

$$S_T = S_{298.15} + \int_{298.15}^{T} (C_p/T) dT + \sum H_{trs}/T_{trs}, \tag{7.20}$$

where $S_{298.15}$ is the standard entropy of the substance, which can be calculated by integrating C_p/T functions from 0 to 298.15 K, T is temperature, and H_{trs} is enthalpy of phase transition at a temperature T_{trs}.

From HSC Chemistry 6.0 μ^0 values at 1700 K are derived from Equations 6.1, 7.19, and 7.20. $\mu^0_{FeO} = -451.193$ kJ mol^{-1}, $\mu^0_{CO} = -494.471$ kJ mol^{-1}, $\mu^0_{Fe} = -98.124$ kJ mol^{-1}, $\mu^0_{CO_2} = -829.283$ kJ mol^{-1} and $RT = 14,134$ J mol^{-1}.

$$G(\xi) = -451,193(1-\xi) + (1-\xi)\{-494,471 + 14,134\ln(1-\xi)\}$$
$$= -98,124\xi + \xi\{-829,283 + 14,134\ln\xi\}. \tag{7.21}$$

Figure 7.8 shows $G(\xi)$ plotted against extent of reaction ξ.

If we want to determine ξ_{eq}, we must differentiate $G(\xi)$ with respect to ξ. With Mathcad* we can derive

---

* Mathcad is a desktop software program for performing and documenting engineering and scientific calculations. Mathcad was conceived and originally written by Allen Razdow (of MIT), co-founder of Mathsoft, which is now part of Parametric Technology Corporation.

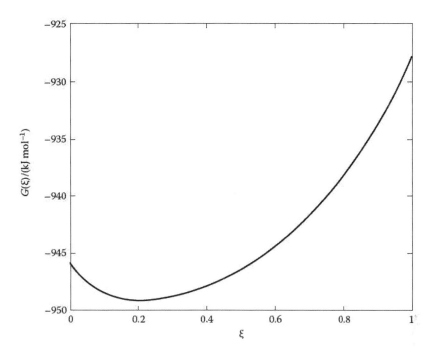

FIGURE 7.8 The Gibbs free energy, $G(\xi)$, of the reaction mixture as a function of the reaction ξ for $FeO(1) + CO(g) \rightleftarrows Fe(s) + CO_2(g)$ at 1700 K and 1 bar.

$$\left(\frac{\partial G}{\partial \xi}\right)_{T,P} = 14{,}134 \ln \xi - 14{,}134 \ln (1-\xi) + 18{,}257.$$

At equilibrium, $\left(\dfrac{\partial G}{\partial \xi}\right)_{T,P} = 0$.

With Mathcad or a scientific pocket calculator we can calculate that $\xi_{eq} = 0.216$,

$$K_p = \frac{P_{CO_2}}{P_{CO}} = \frac{x_{CO_2} P_{tot}}{x_{CO} P_{tot}} = \frac{\xi}{1-\varepsilon} = \frac{0.216}{1-0.216} = 0.28,$$

which is in agreement with K_p calculated from ΔG^0.

In some other publications [2,3], in this kind of calculation, the standard state is chosen by ignoring the μ^0 values for the elements from which they are made. This is clearly not conceptually correct, but in turn shifts the graph of G_{system} versus ξ up or down on the y-axis but does not affect their shape and does not affect the results. We must remember that:

$$\Delta_f G^0 = \mu^0_{compound} - \sum (\text{moles elements}) \mu^0_{element}.$$

The effect of excluding the $\mu^0_{elements}$ on G_{system} calculated from Equation 7.16 is simply to subtract a constant that is independent of the extent of reaction ξ. Therefore, ignoring the $\mu^0_{elements}$ removes that constant from the value of G_{system}. In problem 7.12 the student should check this.

7.8 SUMMARY

We have derived a deceptively simple equation, $\Delta G^0 = -RT \ln K$, that enables us to calculate equilibrium constants from ΔG^0 data.

ΔG^0 determines how far the reaction will proceed before reaching equilibrium, not whether the reaction is initially spontaneous. Also from a graph of G for a reaction mixture versus the extent of reaction ξ, ξ_{eq} and the equilibrium constant K can be calculated.

Having derived and discussed the concept of activity, we can easily take into account the nonideal behavior of solutions and of gases at significant pressures. However, in cases where solutions are dilute, or gas pressures low, calculations for the ideal case are also easily made by putting all activity coefficients equal to unity.

REFERENCES

1. Clerc, D. G., and D. A. Cleary, 1995. Spinodal decomposition as an interesting example of the application of several thermodynamic principles. *J Chem Ed* 72:112–115.
2. Ferguson, A. 2004. The Gibbs free energy of a chemical reaction system as a function of the extent of reaction and the prediction of spontaneity. *J Chem Ed* 81:606.
3. McQuarrie, D. A., and J. D. Simons, 1999. *Molecular Thermodynamics*. University Science Books: Sausalito, California, 486.

PROBLEMS

7.1   At 2100 K, the equilibrium constant for the reaction:

$$Si(l) + 2H_2(g) \rightleftarrows SiH_4(g)$$

is $K = 9.17 \times 10^{-7}$.
(a) What is the ΔG^0 for this reaction?
(b) If ΔS^0_{2100} is -127.1 J mol^{-1} K^{-1}, what is the enthalpy change?

7.2   An experimental study of cobaltous silicate, Co_2SiO_4, is launched. In a two-pronged approach, the carbon monoxide reduction of Co_2SiO_4 and of CoO is studied. For these reactions,

$$Co_2SiO_4(s) + 2CO \rightarrow 2CO_2 + 2Co(s) + SiO_2(s) \qquad I$$

$$2CoO(s) + 2CO \rightarrow 2CO_2 + 2Co(s) \qquad II$$

at 1300 K, the CO_2/CO pressure ratios were 4.30 for I and 12.76 for II.
(a) Calculate ΔG^0_{1300} in each case. (*Hint:* Don't forget about the coefficients in the reaction equation.)
(b) By combining these data suitably, find ΔG^0_{1300} for the combination of oxides:

$$2CoO + SiO_2 \rightarrow Co_2SiO_4.$$

7.3 For approximate determinations, ΔG_T^0 may be calculated as

$$\Delta G_T^0 = \Delta H_{298}^0 - T\Delta S_{298}^0.$$

Use tabulated data to determine an equilibrium constant for one of the following industrial processes. (*Hint:* Use both $\Delta_f G_{298}^0$ and $\Delta_f H_{298}^0$ to find $\Delta_f S_{298}^0$, rather than deriving from individual S^0 data.)

(a) Manufacture of CS_2:

$$CH_4(g) + 2S_2(g) \xrightarrow[650°C]{\text{Bauxite}} 2H_2S(g) + CS_2(g).$$

(b) Production of HCN:

$$CH_4 + NH_3 + \tfrac{3}{2}O_2 \xrightarrow[1000°C]{\text{Pt/Rh}} HCN + 3H_2O(g).$$

(c) Acetylene, by the Sachse process:

$$2CH_4 \underset{\text{(natural gas)}}{\xrightarrow{1500°C}} C_2H_2 + 3H_2 \begin{pmatrix} \text{Rapid product cooling} \\ \text{to avoid back reaction} \end{pmatrix}$$

7.4 In a study of the reaction:

$$CO_2(g) + C(\text{graphite}) \rightleftarrows 2CO(g),$$

at 796°C, it was found that at a total pressure of 1.0460 bar the partial pressure of CO was 0.9187 bar. Calculate the equilibrium constant for this reaction, and thence a value of ΔG^0 at this temperature.

7.5 The standard free energy of formation of N_2O_4 at 25°C is 99.80 kJ mol^{-1} and that of NO_2 is 52.32 kJ mol^{-1}. Calculate the equilibrium constant for this reaction and the fraction of N_2O_4 that will be dissociated if a sample of the gas is allowed to attain equilibrium at 25°C and a total pressure of 1 bar.

7.6 Tin of high purity is to be cast at a temperature of 600 K, at which ΔG^0 for the reaction:

$$Sn(l) + O_2(g) \rightarrow SnO_2(s)$$

is -452.4 kJ mol^{-1}.

Argon is used as a blanketing gas, but usually contains oxygen at a partial pressure of 10^{-6} bar. Will tin oxidize in this environment? (*Hint:* This is not an equilibrium situation.)

7.7 For the reaction:

$$ZnSO_4 \cdot 7H_2O \rightarrow ZnSO_4 \cdot 6H_2O + H_2O(l),$$

$\Delta G^0 = 2.916$ kJ mol^{-1} at 18°C. The vapor pressure of water is 15.48 mm Hg at the same temperature. What is the vapor pressure of zinc sulfate heptahydrate in equilibrium with the hexahydrate at 18°C?

(*Hint:* The water vapor pressure gives the free energy change for the vaporization process and sets up a reaction scheme as in Example 2.10.)

7.8   Use tabulated data to find the partial pressure of fluorine atoms at equilibrium in fluorine gas at 1 bar and at 298.15 K.

7.9   The computer program HSC Chemistry was used to calculate K for the reaction:

$$CO + 2H_2 \rightleftarrows CH_3OH(g),$$

as 2.11×10^{-3} at 250°C. Calculate ΔG^0 for this reaction at this temperature. Using tabulated data, find ΔG^0 at 298 K. Why is it necessary in practice to carry out this methanol synthesis at high temperature and pressure rather than at 298 K?

7.10   On heating, nitrosyl chloride at 650 K NOCl decomposes into nitric oxide and chlorine:

$$2NOCl(g) \rightarrow 2NO(g) + Cl_2(g).$$

The reaction is endothermic, with $\Delta H^0 = 77.095$ kJ mol^{-1} Cl$_2$. If it may be assumed that ΔH and ΔS are temperature invariant, and the standard molar entropies for NOCl, NO, Cl$_2$ are 299.033, 234.394, and 250.802 J mol^{-1} K^{-1}, respectively, calculate the equilibrium constant for the reaction at 650 K.

7.11   Consider the reaction of Problem 7.10:

$$2NOCl(g) \rightarrow 2NO(g) + Cl_2(g)$$

at 650 K and a total pressure of 1 bar. Suppose we start with 2 mol of NOCl(g), and no NO(g) or Cl$_2$(g).

(a) Calculate the Gibbs free energy of the reaction mixture, G, as a function of the extent of the reaction ξ, using $\mu^0(NOCl) = -125.611$ kJ mol^{-1}, $\mu^0(NO) = -51.336$ kJ mol^{-1}, and $\mu^0(Cl_2) = -150.446$ kJ mol^{-1}.

(b) Draw a plot of $G(\xi)$ against ξ.

(c) With a computer program such as Mathcad or a scientific pocket calculator differentiate $G(\xi)$ with respect to ξ and find

$$\left(\frac{\partial G}{\partial \xi} \right)_{T,P}.$$

(d) Calculate ξ_{eq} and the equilibrium constant K_p.

7.12   In Example 7.3, we studied the reaction:

$$FeO(l) + CO(g) \rightleftarrows Fe(s) + CO_2(g),$$

at 1700 K and a total pressure of 1 bar. Suppose we start with 1 mol FeO(l) and 1 mol CO(g):

(a) Calculate the Gibbs free energy of the reaction mixture, G, as a function of the extent of the reaction ξ by setting the values for $\mu^0_{compound}$ equal to their $\Delta_f G^0$ values, ignoring the μ^0 values for the elements. $\Delta_f G^0$ values, at 1700 K calculated with HSC Chemistry are: $\Delta_f G^0(FeO) = -153.770$ kJ mol^{-1}, $\Delta_f G^0(CO) = -260.796$ kJ mol^{-1}, $\Delta_f G^0(\text{Fe}) = 0$ kJ mol^{-1}, and $\Delta_f G^0(\text{CO}_2) = -396.309$ kJ mol^{-1}.

(b) Draw a plot of $G(\xi)$ against ξ.

(c) With a computer program such as Mathcad or a scientific pocket calculator, differentiate $G(\xi)$ with respect to ξ and find

$$\left(\frac{\partial G}{\partial \xi} \right)_{T,P}.$$

(d) Calculate ξ_{eq} and the equilibrium constant K_p.

(e) If you compare your results with that from Example 7.4, which conclusions can you draw?

8 Equilibrium Experiments and Their Interpretation

In this chapter, we shall derive one further equation, the reaction isochore (sometimes called "isobar"), and apply this and other equations to a variety of experiments. We shall see that relatively simple measurements of equilibria, taken over a range of temperature, provide information on both enthalpy changes, ΔH, and entropy changes, ΔS, and therefore on changes of free energy, ΔG.

8.1 THE REACTION ISOCHORE EQUATION

This equation is often linked with the name of van't Hoff, who was the first to apply it widely. It may be derived in several ways. We shall adopt the simplest method, which will make full use of the concept of activity. We start with Equations 6.3 and 7.15, which should by now be engraved indelibly on the minds of all readers. They are:

$$\Delta G = \Delta H - T\Delta S \quad (T \text{ constant}), \tag{6.3}$$

which is, for standard conditions:

$$\Delta G^0 = \Delta H^0 - T\Delta S^0 \quad (T \text{ constant}) \tag{8.1}$$

and

$$\Delta G^0 = -RT \ln K. \tag{7.15}$$

If ΔG^0 is eliminated, we obtain:

$$\ln K = -\frac{\Delta H^0}{RT} + \frac{\Delta S^0}{R} \quad (T \text{ constant}) \tag{8.2}$$

This is the van't Hoff isochore (or isobar) equation. We shall use this equation often, although at first sight, it seems not to be very helpful. Strictly speaking, there are *four* variables, K, ΔH^0, T, and ΔS^0, although only K varies widely with temperature. However, it is worthwhile to look more closely at ΔH^0 and ΔS^0 and their variation with temperature.

From Equation 3.4 we know that the variation in ΔH is given by:

$$\frac{d(\Delta H)}{dT} = \Delta C_p.$$

When integrated, this equation states that:

$$\Delta H_2 - \Delta H_1 = \int_{T_1}^{T_2} \Delta C_p dT.$$

Thus, the difference in heat capacities across the reaction determines the change of ΔH from T_1 to T_2. Usually, the heat capacities of the products almost equal those of the reactants, and ΔC_p is very small indeed. Often, ΔH is assigned a constant value. Over large ranges of temperature, the integral must be evaluated exactly; we saw to this in Chapter 3.

The argument for changes in ΔS runs on parallel lines. For a single material, because $dS = dq/T$:

$$\int_{T_1}^{T_2} dS = \int_{T_1}^{T_2} \frac{C_p dT}{T},$$

and so

$$S_2 - S_1 = \int_{T_1}^{T_2} \frac{C_p dT}{T}.$$

For a reaction, we are concerned with ΔC_p, and so:

$$S_2 - S_1 = \int_{T_1}^{T_2} \frac{\Delta C_p dT}{T}.$$

As before, ΔC_p is often very small, and this equation need be calculated only for large changes in temperature. (Note that although ΔS^0 may be temperature-invariant, absolute entropies are certainly not.)

Let us return to the isochore equation (Equation 8.2). To recapitulate, we have shown that ΔH^0 and ΔS^0 may be regarded as constant over small ranges of temperature, and so the equation will simply show the variation of K with temperature. Let us therefore modify it, and write it simply as:

$$\ln K = -\frac{\Delta H^0}{R}\frac{1}{T} + C \quad (T \text{ constant}), \tag{8.3}$$

where C is a constant, equal to $\Delta S^0/R$. Notice that the restriction on constant temperature applies only to a particular experiment, which lasts only long enough to take the necessary measurements. Further experiments at different temperatures would be quite possible. We should find that K would vary exponentially with $1/T$ if Equation 8.3 is valid. Further, if we recognize that Equation 8.3 has the classic form of the $y = mx + c$ relationship, we see that y corresponds to $\ln K$, x to $1/T$, and m, the gradient, to $-\Delta H^0/R$, that is, the slope of the straight-line graph:

$$\frac{d(\ln K)}{d(1/T)} = -\frac{\Delta H^0}{R}.$$

We shall use this relationship later in the chapter.

It is possible, if data are scarce, to obtain meaningful results with only two data, preferably spaced well apart. Equation 8.3 must hold for both sets of values, 1 and 2, and we can write:

$$\ln K_1 = -\frac{\Delta H^0}{R}\frac{1}{T_1} + C$$

and

$$\ln K_2 = -\frac{\Delta H^0}{R}\frac{1}{T_2} + C.$$

On subtracting these, we have:

$$\ln\frac{K_1}{K_2} = -\frac{\Delta H^0}{R}\left(\frac{1}{T_1} - \frac{1}{T_2}\right). \tag{8.4}$$

This is a useful form of the equation for purposes of simple calculation.

One final point should be made concerning equilibria in condensed phases (no gases present). In such a case, K would be expressed in terms of concentrations or mole fractions, and we would be justified in putting K_c in its place. In addition, ΔH need no longer be used; ΔU may be substituted, because both pressure *and* volume will remain constant. A very simple proof would look like this:

Because $H = U + PV$, therefore $\Delta H = \Delta U + (P_2V_2 - P_1V_1)$,

and, because P and V are both constant (condensed phases):

$$\Delta H = \Delta U.$$

This isochore will now wear a very thin disguise, but it will be recognizable nonetheless, as:

$$\ln K_c = -\frac{\Delta U^0}{RT} + \frac{S^0}{R} \qquad (P, V \text{ constant}).$$

8.1.1 LE CHATELIER UP TO DATE

Because K is taken to be a function of temperature only, we may differentiate $\ln K$ with respect to T. Thus, if we write the isochore as:

$$\ln K = -\frac{\Delta H^0}{R}T^{-1} + C,$$

$$\frac{d(\ln K)}{dT} = \frac{\Delta H^0}{RT^2}. \tag{8.5}$$

Let us consider a particular reaction. Iodine vapor dissociates into atoms to a progressively greater extent as the temperature is raised. Is this fact amenable to thermodynamic treatment? This equation shows us that it is. Looking at it qualitatively, we see that the sign of the right-hand side is the same as that of ΔH. For the reaction

$$I_2(g) \rightleftarrows 2I(g),$$

$\Delta H = +115.11$ kJ mol^{-1}, and so $\Delta H/RT^2$ is positive. This means that d($\ln K$)/dT is also positive, and so K increases logarithmically as T increases. Not only can we see this qualitatively, but if we are given K at one temperature, we can also make quantitative estimates of K at other temperatures. Thus, by using Equation 8.5, we see that at 300 K:

$$d(\ln K)/dT = 115{,}110/8.314(300)^2 = 0.15.$$

That is, for every 10 degrees, $\ln K$ changes by 1.5. This means that K increases by *five times every 10 degrees*, in the region of 300 K. This change seems quite sensational but such effects are not uncommon.

This calculation is seen to be in the nature of a quantitative Le Chatelier prediction. He was able to predict the direction of changes in equilibrium with changes of temperature ("such that the effect of the change of conditions shall be minimized") [1], but we are now able to measure its magnitude. This we shall do in the examples that follow, where we shall use the integrated form of the isochore, given as Equation 8.2. The examples demonstrate different experimental techniques, and cover different types of process. The basic thermodynamic data so obtained can be transferred and modified in order to predict and understand new reactions and new processes.

8.2 APPLICATIONS OF THE ISOCHORE EQUATION

We shall discuss three equilibrium studies. New aspects of the isochore will be brought out as they arise.

8.2.1 Vaporization Processes

The equilibrium vapor pressure of liquid gallium over a range of temperatures has been measured and is summarized in Table 8.1.

The Knudsen effusion technique entails heating a small quantity of the material under study in a refractory container, in a high vacuum. Vapor effuses, through a small pinhole in the lid, and sprays onto a cold surface, where it condenses. After several hours, the deposit is weighed. Vapor pressures, such as those in Table 8.1, may then be calculated from the pinhole size, the deposition time, the weight deposited, and other constant factors. If the pinhole is sufficiently small, then equilibrium conditions may be assumed to prevail inside the cell.

The equilibrium process may be represented as:

$$Ga(l) \rightleftarrows Ga(g)$$

$$a = 1 \quad f \approx p.$$

The activity of the liquid is taken as unity, and for the vapor it is valid to assume that fugacity and vapor pressure are equal. Then, the equilibrium constant K, which is $a_{Ga(g)}/a_{Ga(l)}$, becomes equal to p, the vapor pressure. Moreover, enthalpy change is $\Delta_{vap}H$ for the vaporization process. Hence, Equation 8.2 becomes:

$$\ln p(\text{bar}) = -\frac{\Delta_{vap}H}{R}\frac{1}{T} + C. \tag{8.6}$$

TABLE 8.1
Vapor Pressure of Liquid Gallium in the Knudsen Effusion Technique, from 1281 to 1437 K

Temperature (K)	Vapor Pressure (bar)	1/T (T in K)	ln p (p in bar/1 bar)
1281	7.09×10^{-6}	7.81×10^{-4}	−11.856
1311	1.21×10^{-5}	7.63×10^{-4}	−11.326
1330	1.81×10^{-5}	7.52×10^{-4}	−10.918
1355	2.63×10^{-5}	7.38×10^{-4}	−10.544
1368	2.86×10^{-5}	7.31×10^{-4}	−10.463
1373	3.64×10^{-5}	7.28×10^{-4}	−10.222
1390	4.42×10^{-5}	7.19×10^{-4}	−10.027
1413	7.55×10^{-5}	7.08×10^{-4}	−9.492
1437	1.01×10^{-4}	6.96×10^{-4}	−9.197

Source: Alcock et al., *Proc. Conf. On Thermodynamics*, International Atomic Energy Authority, Vienna, 1966. With permission.

Values of ln p and $1/T$ have been calculated, and also appear in Table 8.1. Figure 8.1 shows a graph of ln p versus $1/T$, which gives a good straight line. The gradient is near −31,288 K, and so:

$$\Delta_{vap}H = -R(-31,288) = +260.1 \text{ kJ mol}^{-1}.$$

When sample values of p and T are put into the equation, we find the constant term C to be 12.551. We previously put $C = \Delta S/R$, and so:

$$\Delta S = CR = R(12.551) = 104.4 \text{ J mol}^{-1} \text{ K}^{-1}.$$

This value is the entropy change on converting 1 mol of liquid gallium at about 1400 K to vapor at the same temperature and a pressure of approximately 5 bar. Strictly speaking, because ΔH will have a slightly different value at the boiling point, one should be cautious when comparing this value with Trouton's constant. Nonetheless, the agreement between the indicated value 104.4 and values for mercury (94.1), cadmium (103.8), and zinc (105.4) is encouraging.

As a final check on the vapor pressure equation as it now stands:

$$\ln p(\text{bar}) = -\frac{260,144}{R}\frac{1}{T} + 12.551,$$

we can calculate the boiling temperature, where $p = 1$ atm (= 1.01325 bar). From

$$\ln 1.01325 = -\frac{260,144}{8.3145}\frac{1}{T} + 12.551 \rightarrow T = 2495 \text{ K}.$$

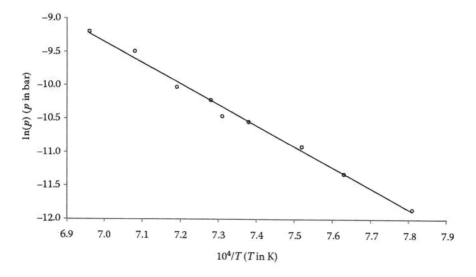

FIGURE 8.1 Vapor pressure of gallium, expressed as ln p versus $10^4/T$ graph, from 1280 to 1400 K.

The Thermal and Physical Properties database (1994) from ES Microware (Hamilton, OH) gives a value of 2478 K. With HSC Chemistry, a value of 2475 K has been estimated.

The vapor pressures of most liquids conform to the simple relationships discussed above. To demonstrate that this is so for high-boiling metals as well as for lower-boiling organic materials, a number of plots of log (vapor pressure) versus $1/T$ are shown in Figure 8.2. They are clearly members of a family of lines, which seem to converge at a point where $1/T = 0$. Although this has little physical meaning, it can be interpreted algebraically as the point where $C = \ln K = \Delta S/R$. From such a series of graphs, C is found to vary from about 10 to 15, corresponding to values of ΔS between 80 and 125 J mol^{-1} K^{-1}. This wide variation only serves to underline the fact that Trouton's rule is at best a rule of thumb.

8.2.2 THE DECOMPOSITION OF THE COMPOUND FE(OH)$_3$

Cheaply produced, clean alternative fuels are much in demand; hydrogen is a promising candidate. The decomposition of Fe(OH)$_3$(s) has been suggested as part of a thermochemical hydrogen production cycle. For the reaction:

$$6Fe(OH)_3(s) \rightleftharpoons 2Fe_3O_4(s) + 9H_2O(g) + \tfrac{1}{2}O_2(g),$$

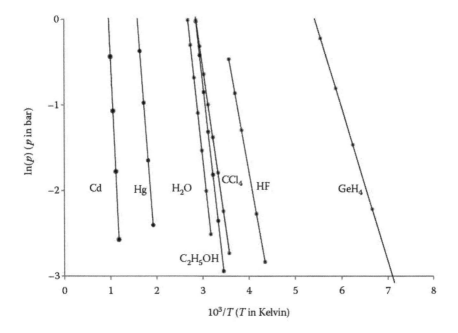

FIGURE 8.2 Graphs of ln p (p in bar) versus $1/T$ (T in Kelvin) for vaporization of various materials.

the data in Table 8.2 have been calculated with the computer program HSC Chemistry 6.1. The equilibrium constant is $K \approx p_{H_2O}^0 \times p_{O_2}^{1/2}$, because the activities of $Fe(OH)_3(s)$ and $Fe_3O_4(s)$ can be taken as 1.

First, we shall derive the Gibbs free energy change as a function of temperature. With the linear regression function on a pocket calculator or PC, $1/T$ (K^{-1}) is taken as the x variable and $\ln K$ as the y variable. This gives a slope of -4.926×10^4 (K) and an intercept of 145.8. The correlation coefficient is -0.998, which means that a plot of $\ln K$ against $1/T$ yields a straight line with a very good fit of the data. The slope and the intercept gives $\Delta H^0 = 409.581$ kJ/(6 mol $Fe(OH)_3$ K) and $\Delta S^0 = 1212$ J/(6 mol $Fe(OH)_3$ K). That is,

$$\ln K = -\frac{409,581}{RT} + \frac{1212}{R}.$$

We have sufficient data to express the Gibbs free energy change of the reaction as:

$$\Delta G^0 = 409,581 - 1212T.$$

From the tabulated data, it seems that the reaction should proceed spontaneously; however, it does not [2]. This behavior can be explained when we realize that the reaction represents the sum of the following reactions:

$$6Fe(OH)_3(s) \rightleftharpoons 3Fe_2O_3(s) + 9H_2O(g) \tag{I}$$

$$3Fe_2O_3(s) \rightleftharpoons 2Fe_3O_4(s) + \tfrac{1}{2}O_2(g) \tag{II}$$

Let us investigate the situation at a temperature of 1400 K. The calculated ΔS^0 values for reactions I and II are:

$$\Delta G^0_{I,1400} = -1330.78 \text{ kJ},$$

TABLE 8.2

Equilibrium Data of the $Fe(OH)_3$ Decomposition Reaction, Calculated with the Computer Program HSC Chemistry 6.1

Temperature (K)	ΔG^0 (kJ)	$\ln K$
1000	−797.94	95.970
1100	−924.28	101.059
1200	−1048.55	105.092
1300	−1170.85	108.323
1400	−1291.32	110.935
1500	−1410.10	113.064
1600	−1527.34	114.810
1700	−1643.16	116.250

---

and

$$\Delta G^0_{\text{II},1400} = +39.46 \text{ kJ}.$$

For reaction I, the equilibrium partial pressure of water is:

$$K_{\text{I}} = 4.53 \times 10^{49} = P^9_{\text{H}_2\text{O}},$$

and so

$$P_{\text{H}_2\text{O}} = 3.3 \times 10^5 \text{ bar}.$$

For reaction II, the equilibrium partial pressure of oxygen is:

$$K_{\text{II}} = 0.0337 = P^{1/2}_{\text{O}_2}$$

and so

$$P_{\text{O}_2} = 1.1 \times 10^{-3} \text{ bar}.$$

For water, partial pressures of less than 3.3×10^5 bar $Fe_2O_3(s)$ will form from the decomposition of $Fe(OH)_3$; hence, there is no problem with reaction I. However, the partial pressure of oxygen must be kept less than 1.1×10^{-3} bar in order to decompose Fe_2O_3 to Fe_3O_4. In air, the partial pressure of oxygen is about 0.2 bar. This means that the decomposition of Fe_2O_3 to Fe_3O_4 can occur only in a vacuum or when a large flow of inert gas is maintained. This is the reason why $Fe(OH)_3$ does not decompose under normal conditions at the temperatures given in Table 8.2.

8.2.3 THE HIGH-TEMPERATURE DISSOCIATION OF WATER VAPOR

As a final example, we shall consider the origins of the data on that most important of species, the water molecule. Such quantities as ΔG^0 and ΔH^0 for the formation of water are used so frequently that it is essential to have reliable, accurate data to draw upon. For example, water dissociation equilibria will be an important factor in flame and rocket technology. Such processes are strongly endothermic, and will run away with the available energy. This will limit the maximum temperature of the flame, as pointed out in Chapter 3.

In this example, the equilibrium data depend on determining not one but three separate partial pressures, and more sophisticated tools, such as ultraviolet absorption and emission spectrometers were used to determine the equilibrium constant K at different temperatures. One of the reactions studied was:

$$2H_2 + O_2 \rightleftarrows 2H_2O,$$

for which

$$K = \frac{a_{H_2O}^2}{a_{H_2}^2 a_{O_2}} = \frac{p_{H_2O}^2}{p_{H_2}^2 p_{O_2}}.$$

The pressures were low at all times, and so $\gamma = 1$ throughout. The dissociation is very slight, and so the water vapor pressure is virtually equal to the total vapor pressure. Table 8.3 gives a few of the values of K calculated from such experimental results.

It is most striking to see such vast changes occur in the equilibrium constant over a few hundred degrees. For example, from the temperature of boiling water to that of molten lead, 327°C, K decreases by a factor of more than 10^{25}. However, even such changes as these are amenable to the isochore equation. It is suggested that the student should draw his own graph to verify the results of the calculations that follow. From the gradient and a spot value on the graph of $\ln K$ versus $1/T$ K, we obtain the equation:

$$\ln K = +\frac{59,138}{T} - 12.924.$$

This quickly yields values for:

$$\Delta H^0 = -(8.314)(59.138) = -491.7 \text{ kJ mol}^{-1}$$

and

$$\Delta S^0 = -(8.314)(-12.924) = -107.5 \text{ J mol}^{-1} \text{ K}^{-1}.$$

TABLE 8.3
Equilibrium Constants at Various Temperatures for the Formation of 2 Mol of Water Vapor from Hydrogen and Oxygen

T (K)	K	$\ln K$
400	3.03×10^{58}	134.66
500	5.95×10^{45}	105.40
600	1.87×10^{37}	85.82
800	3.87×10^{26}	61.22
1000	1.37×10^{20}	46.37
1200	6.49×10^{15}	36.41
1500	2.93×10^{11}	26.40
2000	1.26×10^7	16.35

As the quoted reaction is twice the equation for the formation of water vapor, which would be:

$$H_2(g) + \tfrac{1}{2}O_2(g) \rightleftharpoons H_2O(g),$$

a value for $\Delta_f H^0_{298}$ of $-491.7/2 = -245.9$ kJ mol^{-1} is given. This is close to the tabulated value in Appendix III. The standard free energy change for the reaction is easily found:

$$\Delta G^0 = \Delta H^0 - T\Delta S^0$$

$$= -491.7 - 298.15(-107.5)/1000$$

$$= -459.6 \text{ kJ mol}^{-1}.$$

From this, $\Delta_f G^0_{298}(H_2O) = -459.6/2 = -229.8$ kJ mol^{-1}. This compares well with the currently accepted value of -228.6 kJ mol^{-1}.

8.3 THE CLAPEYRON EQUATION

Section 8.2.1 was concerned with equilibrium between a condensed phase and the vapor. It is often necessary, however, to estimate the effect of pressure on equilibria between two condensed phases. For example, the melting point of sodium at 1 atm pressure is 97.6°C. Can it be used as a liquid heat transfer medium at 100°C, at a pressure of 100 atm, or will it solidify? It is known that the liquid is less dense than the solid and this argues that high pressures will encourage solidification. This is another aspect of Le Chatelier's work, which we can now quantify. This and similar problems may be solved by the Clapeyron equation, which we shall now derive.

Let us continue with the sodium example, representing the process as Na(s) $\rightleftharpoons$ Na(l). Because solid and liquid sodium are in equilibrium at 97.6°C and 1 atm, their chemical potentials must be equal (i.e., $\mu_s = \mu_l$). Suppose that we can now arrange a different pressure and temperature that allow equilibrium to occur once again. Although the chemical potentials will have changed, they will have done so *by the same amount*; this is necessary to maintain the new potentials equal, that is $\mu_s' = \mu_l'$. Therefore, it follows that $d\mu_s = d\mu_l$.

We shall now use Equation 7.1, expressed in molar quantities. Applied to the solid, it tells us that $d\mu_s = V_m(s)dP - S_m(s)dT$. Also, for the liquid, $d\mu_l = V_m(l)dP - S_m(l)dT$. Because

$$d\mu_s = d\mu_l,$$

$$V_m(s)dP - S_m(s)dT = V_m(l)dP - S_m(l)dT$$

or

$$\frac{dP}{dT} = \frac{\Delta H}{T\Delta V_m}.$$

Finally, our equilibrium situation requires that $\Delta G = \Delta H - T\Delta S = 0$, and so $\Delta S = \Delta H/T$. We can now write the Clapeyron equation as

$$\frac{dP}{dT} = \frac{\Delta H}{T\Delta V_m}. \tag{8.7}$$

The sodium problem may now be tackled. We find the following data:

Heat of fusion = 3.05 kJ mol^{-1},

Density of liquid = 0.9288 g ml^{-1},

Density of solid = 0.9519 g ml^{-1}.

The molecular weight of Na = 23.0 g mol^{-1}. Thus,

$\Delta_{fus}H = 30.1$ L atm mol^{-1} (1 joule = 0.00987 L atm)

$\Delta V_m = +6.05 \times 10^{-4}$ L mol^{-1}.

From Equation 8.7, we calculate:

$$\frac{dP}{dT} = \frac{30.1}{370.8(6.05 \times 10^{-4})} = 134 \text{ atm K}^{-1}.$$

A 1°C rise in melting point to 97.6 + 1 = 98.6°C is accompanied by an increase in pressure of 134 atm, and so use of the liquid sodium at the suggested 100 atm and 100°C would give no problems. Liquid sodium is, in fact, used as a coolant in some types of nuclear reactors.

8.4   SUMMARY

We have seen how data from experimental studies of equilibria are converted into a readily usable form, that is, into ΔH, ΔS, and ΔG values. These data are combined with others, determined, for example, by reaction calorimetry (ΔH) or by the third law (as opposed to the second law) determinations of entropy, using low-temperature heat capacity measurements. Before we can enter the first division of predictive thermodynamics and confidently plan processes, we must learn to accept information from two additional sources. First, there is a great reservoir of electrochemical expertise that we have not yet tapped. Second, we must learn to handle the refinements in free energy formulations, which make allowance for the slight variations of ΔH and ΔS with temperature. The example in Section 8.2.2 illustrates the potential difficulties that can arise through the superficial application of thermodynamics. In this specific application, knowledge of the existence of Fe_2O_3, a species not included in the original reaction, was required.

REFERENCES

1. Le Chatelier, H. L. 1888. Recherches expérimentales et théoriques sur les équilibres chimiques. *Annales des Mines*, 13, 157–382.
2. Bamberger, C. E., J. Braunstein, and D. M. Richardson. 1978. Thermochemical production of hydrogen from water. *J Chem Ed* 55:561–564.
3. Berka, L. H., N. K. Kildahl, S. J. Bergin, and D. S. Burns. 1994. Equilibrium vapor pressures: Comparing pressures obtained from using a classical method with those from a laboratory-computer interface. *J Chem Ed* 71:441–445.

PROBLEMS

8.1   In a study of the vapor pressure of liquid methanol [3], students observed the following results:

T (°C)	P (mm Hg)
32.1	173.0
37.1	223.7
40.0	259.1
43.5	304.3
47.0	359.4
50.5	434.6

Determine $\Delta_{vap}H$ and $\Delta_{vap}S$ from these data graphically, with the linear regression function on your pocket calculator or with Excel on your PC.

8.2   The vapor pressure of water at 25°C is 23.76 mm Hg. Use this datum and the normal boiling point of water to derive an equation for the temperature variation of the vapor pressure of water from 25°C to 100°C. A mountaineer encounters an atmospheric pressure of only 485 mm Hg; at what temperature will water boil?

8.3   The vapor pressure of solid and liquid Tc_2O_7 is as follows:

Solid		Liquid	
T (K)	p (mm Hg)	T (K)	p (mm Hg)
391.2	0.65	393.3	1.0
388.7	0.46	413.1	2.1
384.1	0.36	427.5	4.0
377.9	0.18	446.4	9.6
373.2	0.095	459.8	16.7
368.5	0.05	473.8	29.0
362.2	0.03	489.7	54.6

Determine the enthalpies of evaporation and sublimation over the temperature range of study. Graphically, from extrapolation of the solidus and liquidus line, determine the melting point.

8.4 The equilibrium constant K_p for the dissociation of bromine vapor, $Br_2(g) \rightleftarrows 2Br(g)$, changes with temperature as follows:

T (K)	K_p
600	5.32×10^{-12}
800	9.15×10^{-8}
1000	3.27×10^{-5}
1400	2.85×10^{-2}
1600	2.41×10^{-1}

Determine $\Delta_{dis}H^0$ for this reaction and $\Delta_f H^0$ of bromine atoms over this temperature range, given that $\Delta_{vap}H^0 = 29.56$ kJ mol^{-1} Br$_2$.

8.5 A study of the equilibrium $CO_2 + C \rightleftarrows 2CO$ yielded the following results:

T (K)	Total Pressure (bar)	Mol % CO_2 in Equilibrium Mixture
1073	2.57	22.18
1173	2.30	5.65

Determine the equilibrium constant K_p for this reaction at 1073 and 1173 K and, hence, $\Delta_r H^0$ between these temperatures. For the equilibrium $2CO_2 \rightleftarrows 2CO + O_2$, the value of K_p at 1173 K is 8.26×10^{-17}; the heat of combustion of carbon to carbon dioxide at 1173 K is 395.0 kJ mol^{-1}. Calculate ΔH^0 and ΔS^0 for the reaction:

$$2CO_2 \rightleftarrows 2CO + O_2 \text{ at } 1173 \text{ K.}$$

8.6 At 251°C, K_p for the dissociation of PCl$_5$:

$$PCl_5(g) \rightleftarrows PCl_3(g) + Cl_2(g)$$

is found to be 1.176. At 307°C, it has reached 7.561. Calculate ΔG^0 at each temperature, and also the average value of ΔH^0 for this temperature range. Compare these data with current values in Appendix III.

8.7 The following data are available for the reaction $Fe_2O_3(s) + 3CO(g) \rightleftarrows 2Fe(s) + 3CO_2(g)$:

T (°C)	100	400	700	1000	1300
K_p	22,093	250	28.8	9.41	5.97

Determine $\Delta_r G^0$ as a function of temperature from these data for this reaction. At 1120°C for the reaction:

$$2CO_2(g) \rightleftharpoons 2CO(g) + O_2(g)$$

gives $K_p = 7.5 \times 10^{-13}$.

Determine the $\Delta_r G^0$ for this reaction at 1120°C.

What equilibrium partial pressure of O_2 would have to be supplied to a vessel at 1120°C containing solid Fe_2O_3 to prevent the formation of Fe? (*Hint:* Set up a reaction scheme such as in Example 2.10.)

8.8 Boron nitride is a refractory compound of great strength that is finding application as a fiber material. For its formation by the reaction:

$$B(s) + \tfrac{1}{2}N_2(g) \rightarrow BN(s),$$

K_p is 210.19 at 1900 K, but it drops to 47.09 at 2100 K. What is the enthalpy change for formation at 2000 K? This datum may be used to calculate the bond dissociation energy of (B—N) in the crystal, which is of diamond structure. (How many moles of bonds in a mole of BN?) For 2000 K, use the following heats of atomization:

$$\Delta_f H^0_{2000}(B(g)) = 558.5 \text{ kJ mol}^{-1},$$

$$\Delta_f H^0_{2000}(N(g)) = 480.0 \text{ kJ mol}^{-1}.$$

8.9 The ionic product of water, $k_w = [H^+][OH^-]$, varies with temperature. It is only 0.2920×10^{-14} at 10.0°C, but rises to 2.919×10^{-14} at 40.0°C.

Use this information to calculate the enthalpy change of neutralization of a strong acid by a strong base, both being of unit activity. Using this value and standard data, calculate $\Delta_f H^0_{298}(OH^-(aq))$.

8.10 Aluminum expands slightly on melting, and $\Delta V = +1.9 \times 10^{-5} \text{ L g}^{-1}$. The melting point is 660.3°C, and $\Delta_{fus} H = 397 \text{ J g}^{-1}$. A sensitive probe is being developed to determine very high pressures by measuring the melting temperature. A change of 0.001°C can be detected. What change of pressure does this represent ($J = 9.87 \times 10^{-3} \text{ L atm}$)?

8.11 Sulfur melts at 115.21°C under 1 atm, and $\Delta_{fus} H = 53.67 \text{ J g}^{-1}$. Densities are 1.811 g mL^{-1} (liquid) and 2.050 g mL^{-1} (solid). In the Frasch process, molten sulfur is pumped from underground deposits at 6 atm total pressure. What must be its minimum temperature?

9 Electrochemical Cells

Her own mother lived the latter years of her life in the horrible suspicion that electricity was dripping invisibly all over the house.

(James Thurber, *My Life and Hard Times*)

In fact, James Thurber's grandmother was partly right; many of the chemical processes occurring around us involve the movement of charged ions in liquids, as well as of electrons in metals. As examples, many corrosion processes arise when aqueous electrolytes are in contact with iron, steel or copper, and battery-powered appliances are common.

Electrochemical cells give valuable thermodynamic information, as well as being useful in their own right. Measurements of cell electromotive force (e.m.f.) yield free energy changes directly, and tabulations of electrode potentials provide an alternative source of free energy data. Variations of e.m.f. with concentration lead to a better understanding of activities, and we shall also see how to measure entropy changes from the variation of cell e.m.f. with temperature.

9.1 ELECTROCHEMICAL CELLS

A cell is a device for harnessing the energy of a chemical process to do electrical work, *directly*. A cell e.m.f. is available at the electrodes, which are immersed in, or in contact with, the electrolyte in which charged species, ions, are mobile.

If a bar of zinc is dipped into a diluted solution of zinc sulfate, some zinc ions, Zn^{2+}, dissolve, leaving two electrons each on the metal. This causes a separation of charge, and eventually (Figure 9.1) equilibrium is achieved. An electrical double layer forms, which consists of electrons on the metal surface and zinc ions immediately adjacent to it. At this stage, the tendency to dissolve is exactly matched by the tendency of zinc ions to deposit, which is caused by the charge separation. This means that there is a potential difference between metal and solution, which, however, cannot be measured. If the potential difference could be measured, we would have a direct measure of the free energy of formation of hydrated zinc ions:

$$Zn(s) \rightarrow Zn^{2+}(aq) + 2e.$$

However, the second lead to the voltmeter would come from the solution and this would constitute a second electrode. Moreover, reactions of this kind, which entail charge separation, occur to only a minute extent.

So, it is not practicable to measure the potential of one electrode relative to that of the electrolyte, as the potential fall across the *two* metallic conductors suspended in the electrolyte is always determined; it is therefore only possible to express the potential of one electrode *relative* to that of another, and an arbitrary reference electrode is required.

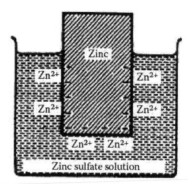

FIGURE 9.1 Ions continue to leave the metal as long as they can overcome the opposing electrostatic forces. Electrons on the zinc attract them, whereas the other zinc ions repel them.

If we can construct a cell from two half-cells such as that described above, the e.m.f. does give free energy information as long as certain rules are obeyed. As an example, consider the Daniell cell, in which a porous barrier separates zinc in zinc sulfate solution from copper in copper sulfate solution. Such a cell may be depicted by the cell diagram:

$$\ominus Zn|ZnSO_4(aq)||CuSO_4(aq)|Cu^{\oplus},$$

where a single line represents the interface between an electrolyte and another phase, and the double line represents the link between the two solutions. The reaction equation, by convention, is written starting from the left, that is, with the Zn electrode:

$$Zn + Cu^{2+}(aq) \rightleftarrows Zn^{2+}(aq) + Cu.$$

The signs at the electrodes represent the charge accumulating at each. Copper is the more noble metal of the two, and cupric ions tend to deposit, making the copper electrode positive. No reaction occurs until these charges are dispersed, usually via some outside circuit in which electrons will flow from the zinc to the copper. The electrode reactions will then be:

$$Zn \rightarrow Zn^{2+} + 2e, \text{ and}$$

$$Cu^{2+} + 2e \rightarrow Cu.$$

Reading from left to right, oxidation occurs in the left half-cell and reduction occurs in the right half-cell. The electrode at which oxidation occurs is called the *anode*, whereas the electrode at which reduction occurs is called the *cathode* (i.e., the anode is negatively charged and the cathode is positive). These signs conflict with negative cathodes and positive anodes in electrolytic cells. In these cells, electrical energy is converted into chemical energy, so the reversal in signs might well be expected.

There are three principal types of electrochemical cells. In an electrolytic cell, a current is passed by an external driving force, causing an otherwise nonspontaneous chemical reaction to proceed. In a galvanic cell, the progress of a spontaneous chemical reaction causes the electric current to flow, doing work on the surroundings. An equilibrium electrochemical cell is at the state between an electrolytic cell and a galvanic cell. The tendency of a spontaneous reaction to push a current through the external circuit must be balanced by an external source of e.m.f. that exactly cancels this tendency. If this counter e.m.f. is increased beyond the equilibrium value the cell becomes an electrolytic cell, and if it is decreased below the equilibrium value the cell becomes a galvanic cell.

We could measure the cell e.m.f. (E) by matching it with an exactly equal, but opposite, e.m.f. from a potentiometer. The cell e.m.f. will be a measure of the free energy for the cell reaction if the electrodes are truly *reversible* and if potentials at the liquid junction are negligible. Reversibility implies that an external e.m.f. infinitesimally greater than E causes the cell reaction to be reversed (compare "external pressure" for gaseous work). This is not true of all cells.

To maintain electrical neutrality in each cell compartment, negative ions move from right to left, within the cell, and positive ions move in the opposite direction. In general, they will have different transport numbers; that is, they move at different rates and so carry different proportions of the total current. This results in a small liquid junction potential, which disturbs the cell e.m.f. reading. (Space allows only a brief account of transport numbers to be included here. For a more complete account, the works of Moore [1] and Potter [2] are recommended.) If, instead of using a porous plate, as in the simple Daniell cell, a salt bridge is used to link the two electrolytes, the junction potential is minimized. A salt bridge is generally a tube of agar jelly containing a solution of either KCl or NH_4NO_3, for which transport numbers of anion and cation are equal. Such a cell is represented as:

$$\ominus Zn|ZnSO_4(aq)|KCl\ bridge|CuSO_4(aq)|Cu\oplus.$$

This cell and the next to be discussed are shown in Figure 9.2. The second cell in Figure 9.2 may be represented by:

$$\ominus Pt,\ H_2(gas)|HCl(aq)|AgCl,\ Ag\oplus.$$

The hydrogen gas electrode comprises a platinized platinum foil with pure hydrogen bubbling over it. If the gas and the acid are at unit activity (1.18 M solution), then it is known as the standard hydrogen electrode (SHE), the potential of which is taken as zero. The silver-silver chloride electrode on the other hand operates as a source of chloride ions. The cell reaction and component half-cell reactions are:

$$\text{Left:} \quad \tfrac{1}{2}H_2 \quad \rightleftarrows \quad H^+ \quad + e$$

$$\underline{\text{Right:} \quad AgCl + e \quad \rightleftarrows \quad Cl^- \quad + Ag}$$

$$\tfrac{1}{2}H_2 + AgCl \rightleftarrows H^+Cl^- + Ag.$$

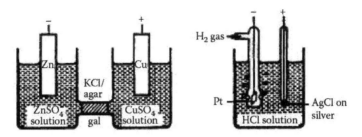

FIGURE 9.2 Two cells discussed in the text. They exemplify the simple metal/ion electrode, the gas/ion electrode, and the metal/metal halide/ion electrode.

Notice that hydrogen appears first, both in the cell diagram and in the cell equation. In this cell, liquid junction potentials do not arise, and if the electrodes are well prepared, reversibility is realized.

The zinc-zinc ion, the hydrogen, and the silver-silver chloride electrodes are typical of three common types of aqueous half-cell. A fourth type is the so-called redox half-cell, which involves, for example, both ferrous and ferric ions in solution. A nonreacting electrode, usually of platinum, facilitates the following half-cell process:

$$Fe^{2+} \rightleftarrows Fe^{3+} + e$$

(from Pt)

The half-cell would be represented as $Pt|Fe^{2+}, Fe^{3+}$; it is basically similar to $M|M^+$ (M = metal), where Fe^{3+} is M^+, and Fe^{2+} is M.

9.2   CELL ENERGETICS

If the reaction

$$Zn(s) + CuSO_4(aq, 0.5\ M) \rightarrow ZnSO_4(aq, 0.5\ M) + Cu(s)$$

is performed in a calorimeter at 15°C, ΔH is found to be −235.2 kJ mol⁻¹. No useful work is done and the process is quite irreversible.

Imagine, however, that the cell:

$$\ominus Zn|ZnSO_4(0.5\ M)|salt\ bridge|CuSO_4(0.5\ M)|Cu\oplus$$

achieves the same overall change at 15°C, but reversibly. Maximum work is done when the load approaches infinite resistance because the cell e.m.f. and the back e.m.f. are then infinitesimally different. This work is measured conveniently by immersing the resistor in a calorimeter. In addition, the cell itself is in another calorimeter to measure q_{rev} of the cell. Figure 9.3 summarizes the situation. The results

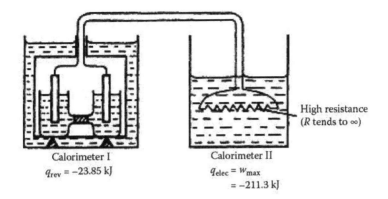

Calorimeter I
$q_{rev} = -23.85$ kJ

Calorimeter II
$q_{elec} = w_{max}$
$= -211.3$ kJ

High resistance
(R tends to ∞)

FIGURE 9.3 In this experiment, the enthalpy of reaction is divided into that available to do work, measured as heat in calorimeter II, and that not available, even during reversible operation (calorimeter I).

show that most of the enthalpy change is available as electrical energy, but 23.85 kJ are lost. Because the operation is reversible, we are already doing our best to maximize the useful work. Also, because $q_{rev}/T = \Delta S$, the term $T\Delta S$ is -23.85 kJ, and $\Delta S = -23,850/288 = -82.8$ J mol^{-1} K^{-1}.

The overall change in enthalpy is the same for each experiment, so:

$$\Delta H = q_{rev} + w_{max} = T\Delta S + w_{max}.$$

(Note that w_{max} represents the work other than that due to gaseous expansion. Gaseous work is already comprehended by ΔH.) Therefore, $w_{max} = \Delta G$, the free energy change for the process, and this equation is seen to be identical with Equation 6.3. We are now able to calculate values of ΔG from simple cell measurements. It is not necessary to perform actual calorimetric experiments. It is often more accurate to calculate the electrical energy from equilibrium cell potentials. As in Section 6.3, the maximum electrical work from the cell is:

$$w_{max} = -\text{charge} \times \text{reversible e.m.f.,}$$

and so

$$\Delta G = -nFE. \tag{9.1}$$

If all reactants and products are in their standard states:

$$\Delta G^0 = -nFE^0. \tag{9.2}$$

Here, n is the number of moles of electrons, or equivalents, involved in the reaction and F, the Faraday constant, is a convenient unit of charge, equal to 96,485 coulombs mol^{-1}.

9.3 STANDARD ELECTRODE POTENTIALS

There are many different half-cells. From these, one can construct many more cells and to tabulate them all would be a huge task. However, we have seen that values of ΔU, ΔH, ΔS, and ΔG for a few simple reactions may be combined to give new values for other reactions. Similarly, data from a few cells may be combined together to yield data for many other possible cells. The cells chosen for tabulation are those involving the SHE, for which the electrode potential is chosen to be at 0 V. What follows is based on the sign convention of the International Union of Pure and Applied Chemistry (IUPAC), agreed to at the Stockholm Convention of 1953. (Please note that other sign conventions have been used and one should take care when consulting books and journals as to which one is being used. Many American books, in particular, use a convention in which opposite signs are used for electrode potentials. Always check the electrode potential of a common reactive metal such as sodium. If it is negative, then the convention is consistent with the IUPAC scheme.)

We adopt the following convention: the standard-state potential difference of a cell consisting of a hydrogen electrode on the left and any other electrode on the right is called the *standard reduction potential* of the right half-cell or the right electrode. It is also sometimes called the *standard electrode potential*.

To write down the diagram Pt, H_2|H^+||Zn^{2+}|Zn necessarily implies a reaction with H_2 as reactant:

$$\tfrac{1}{2}H_2 + \tfrac{1}{2}Zn^{2+} \rightleftharpoons H^+ + \tfrac{1}{2}Zn. \tag{9.3}$$

In fact, for standard states of unit activity, an e.m.f. of -0.763 V is observed, with the zinc negatively charged. The standard-state potential difference of the cell can now be written:

$$E^0(\text{cell}) = E^0(\text{right half-cell}) - E^0(\text{left half-cell})$$

$$= E^0(\text{right half-cell}) - 0$$

$$= -0.763 \text{ V}.$$

This justifies the statement that the standard electrode potential for the Zn^{2+}|Zn half-cell, $E^0(Zn^{2+}|Zn) = -0.763$ V. The sign is that of the charge found on the right-hand side of the cell. Using Equation 9.2,

$$\Delta G^0 = -nFE^0$$

$$= -1(96,485)(-0.763)$$

$$= +73.62 \text{ kJ mol}^{-1} \text{ H}^+ \text{ formed.}$$

This result (ΔG^0 positive) confirms our belief that zinc displaces hydrogen ions from acid solution, rather than vice versa. In the same manner, we interpret the fact that $E^0(\text{Cl}^-|\text{AgCl, Ag}) = +0.2222$ V to imply the following:

(a) $\qquad \ominus\text{Pt, H}_2\,|\text{H}^+||\text{Cl}^-|\,\text{AgCl, Ag}^\oplus$

(b) $\qquad \frac{1}{2}\text{H}_2 + \text{AgCl} \rightleftharpoons \text{H}^+ + \text{Cl}^- + \text{Ag}$

(c) $\qquad \Delta G^0 = -1(96,485)(+0.2222)$

$$= -21.44 \text{ kJ mol}^{-1} \text{ AgCl.}$$

(9.4)

Note that for any half-cell that attains a positive charge when measured against an SHE, the value of ΔG^0 will be negative. This shows that the associated reaction will be spontaneous. Reactions involve transfer of *one* electron for simplicity, and half-cells are written as in the cell diagram.

Consider now a cell in which these two half-cells are combined, such as $\text{Zn}|\text{Zn}^{2+}||\text{Cl}^-|\text{AgCl, Ag}$. The cell reaction is written down, starting as before with the metal (or other reduced species) appearing on the left of the cell diagram (Zn in this case). (In cases where this is not straightforward, remember that product formation in the cell reaction must correspond to movement in the cell of positive charge to the right.) For this cell:

$$\frac{1}{2}\text{Zn} + \text{AgCl} \rightleftharpoons \frac{1}{2}\text{Zn}^{2+} + \text{Cl}^- + \text{Ag.}$$

Then E^0 (cell) is calculated by the right minus left convention of Equation 1.1:

$$E^0(\text{cell}) = E^0(\text{right half-cell}) - E^0(\text{left half-cell})$$

$$= +0.2222 - (-0.763) = +0.9852 \text{ V.}$$

The positive sign is the sign of the charge that turns up spontaneously on the right side of the cell. This means that the cell reaction does, in fact, proceed spontaneously as written. This is confirmed by the negative value of ΔG^0 that now emerges. Using Equation 9.2:

$$\Delta G^0 = -nFE^0 = -1(96,485)(+0.9852)$$

$$= -95.06 \text{ kJ mol}^{-1} \text{ AgCl.}$$

Table 9.1 gives a selection of standard electrode potentials. An excellent collection of oxidation potentials (which are of the opposite sign) has been made by Latimer [3].

TABLE 9.1
A Selection of Standard Electrode Potentials, E^0

Electrode	Cell Reaction	E^0 (V)
Li$^+$ | Li	$\frac{1}{2}H_2 + Li^+ \rightleftharpoons Li + H^+$	−3.025
K$^+$ | K	$\frac{1}{2}H_2 + K^+ \rightleftharpoons K + H^+$	−2.925
Ca^{2+} | Ca	$\frac{1}{2}H_2 + \frac{1}{2}Ca^{2+} \rightleftharpoons \frac{1}{2}Ca + H^+$	−2.87
Na$^+$ | Na	$\frac{1}{2}H_2 + Na^+ \rightleftharpoons Na + H^+$	−2.714
Mg^{2+} | Mg	$\frac{1}{2}H_2 + \frac{1}{2}Mg^{2+} \rightleftharpoons \frac{1}{2}Mg + H^+$	−2.37
Zn^{2+} | Zn	$\frac{1}{2}H_2 + \frac{1}{2}Zn^{2+} \rightleftharpoons \frac{1}{2}Zn + H^+$	−0.763
Fe^{2+} | Fe	$\frac{1}{2}H_2 + \frac{1}{2}Fe^{2+} \rightleftharpoons \frac{1}{2}Fe + H^+$	−0.440
Cd^{2+} | Cd	$\frac{1}{2}H_2 + \frac{1}{2}Cd^{2+} \rightleftharpoons \frac{1}{2}Cd + H^+$	−0.403
Sn^{2+} | Sn	$\frac{1}{2}H_2 + \frac{1}{2}Sn^{2+} \rightleftharpoons \frac{1}{2}Sn + H^+$	−0.136
H$^+$ | H$_2$, Pt		0.00
Cl$^-$ | AgCl, Ag	$\frac{1}{2}H_2 + AgCl \rightleftharpoons Ag + H^+ + Cl^-$	+0.2222
Cl$^-$ | Hg$_2$Cl$_2$, Hg	$\frac{1}{2}H_2 + Hg_2Cl_2 \rightleftharpoons Hg + H^+ + Cl^-$	+0.2673
Cu^{2+} | Cu	$\frac{1}{2}H_2 + \frac{1}{2}Cu^{2+} \rightleftharpoons \frac{1}{2}Cu + H^+$	+0.344
Fe^{3+} | Fe^{2+}, Pt	$\frac{1}{2}H_2 + Fe^{3+} \rightleftharpoons Fe^{2+} + H^+$	+0.771
Ag$^+$ | Ag	$\frac{1}{2}H_2 + Ag^+ \rightleftharpoons Ag + H^+$	+0.799
Br$^-$ | Br$_2$, Pt	$\frac{1}{2}H_2 + \frac{1}{2}Br_2 \rightleftharpoons H^+ + Br^-$	+1.066
Cl$^-$ | Cl$_2$, Pt	$\frac{1}{2}H_2 + \frac{1}{2}Cl_2 \rightleftharpoons H^+ + Cl^-$	+1.359
Ce^{4+} | Ce^{3+}, Pt	$\frac{1}{2}H_2 + \frac{1}{2}Ce^{4+} \rightleftharpoons Ce^{3+} + H^+$	+1.61

In general:
$\frac{1}{2}H_2$ (Oxidized state) $\leftrightarrow$ (Reduced state) + H$^+$

Note: Potentials apply to the cell Pt, H$_2$ | H$^+$ | M$^+$ | M; positive values imply spontaneity, with all materials at unit activity.

Example 9.1

Find the equilibrium constant at 25°C for the reaction:

$$Hg + Cl^- + Ce^{4+} \rightarrow \tfrac{1}{2}Hg_2Cl_2 + Ce^{3+}.$$

This reaction would occur in the cell Hg, Hg$_2$Cl$_2$|Cl$^-$||Ce^{4+},Ce^{3+}|Pt, and from Table 9.1:

$$E^0(\text{cell}) = E^0(\text{right}) - E^0(\text{left})$$
$$= +1.61 - 0.27 = 1.34 \text{ V}.$$

Then

$$\Delta G^0 = -nFE^0$$

$$= -(96,485)(+1.34)$$

$$= -129.3 \text{ kJ mol}^{-1} \text{ Hg}.$$

Furthermore,

$$\Delta G^0 = -RT \ln K$$

whence $K = 10^{23}$.

9.4   VARIATION OF CELL E.M.F. WITH ACTIVITY

We are now in a position to calculate the standard e.m.f. of a cell, E^0, for standard state conditions, but unit activities are seldom encountered. There is a mass action effect, and we shall now investigate this.

The general reaction isotherm is

$$\Delta G = \Delta G^0 = +RT \ln \frac{a_P^p a_Q^q \cdots}{a_A^a a_B^b}. \tag{7.14}$$

If ΔG and ΔG^0 refer to a cell reaction, Equations 9.1 and 9.2 may be enlisted and the activity quotient written Q, to give:

$$E = E^0 - \frac{RT}{nF} \ln Q. \tag{9.5}$$

This important expression is known as the Nernst equation. Now, the reaction equation for cells that include the SHE is:

$$\tfrac{1}{2} H_2 + \text{Oxidized States} \rightleftharpoons \text{Reduced states} + H^+,$$

so

$$E = E^0 - \frac{RT}{nF} \ln \frac{a_{H^+} \cdot a_{\text{Reduced States}}}{a_{H_2}^{1/2} \cdot a_{\text{Oxidized States}}}.$$

However, for cells that include the SHE, $a_{H^+} = a_{H_2} = 1$ and so, in general, for electrode potentials (after inverting the ln term and reversing the sign):

$$E = E^0 + \frac{RT}{nF} \ln \frac{a_{\text{Oxidized States}}}{a_{\text{Reduced States}}}. \tag{9.6}$$

We must remember that the activities appear as they would in the activity quotient. As before, F is the Faraday constant and n is the number of moles of electrons involved in the process. If decadic logarithms are used, and the temperature is 298 K, a value of 0.05916 may be used for $RT(2.303)/F$.

Example 9.2

The following cell is used to determine ion activities in chloride solutions.

$$Pt|Fe^{2+}(a = 0.2),\ Fe^{3+}(a = 0.05)||Na^+Cl^-(a = ?)|AgCl,\ Ag.$$

This would involve two vessels, linked by a salt bridge; in the left, one would be a ferrous-ferric salt solution, with a platinum electrode, and in the right, the sodium chloride solution with a silver-silver chloride electrode. The mean ionic activities (discussed later) are indicated.

E for a particular chloride solution is -0.430 V. What is its activity? The cell reaction is:

$$Fe^{2+} + AgCl \rightleftarrows Fe^{3+} + Cl^- + Ag.$$

(Note that the equation begins with the reduced state (Fe^{2+}) in place of a metal.)
Now,

$$E^0(\text{cell}) = E^0(\text{right}) - E^0(\text{left})$$
$$= 0.2222 - (+0.771)$$
$$= -0.5488\ \text{V}.$$

Using Equation 9.5, we find that:

$$-0.430 = -0.5488 - 0.05916 \log \frac{a_{Fe^{3+}} a_{Cl^-} a_{Ag}}{a_{Fe^{2+}} a_{AgCl}}. \qquad (9.7)$$

The activities of silver and silver chloride, being crystalline solids, are each unity. Further arithmetic shows that:

$$\log \frac{a_{Cl^-}(0.05)}{0.2} = -2.01,$$

whence,

$$a_{Cl^-} = 0.04.$$

Example 9.3

Many sports journalists say that after a rain shower in the summer, long distance runners can run faster because there is more oxygen in the air. It is doubtful that this is true, but e.m.f. measurements could clarify the situation. The measurement of p_{O_2} in the air can be carried out with $ZrO_2(+CaO)$ as the solid electrolyte in the reversible cell:

$$Ni, NiO|ZrO_2(+CaO)|O_2.$$

The electrode reactions are:

$$Ni + O^{2-} \rightarrow NiO + 2e$$
$$\tfrac{1}{2}O_2 + 2e \rightarrow O^{2-}.$$

The net chemical reaction of the cell for the passage of 4 faradays is:

$$2Ni + O_2 \rightarrow 2NiO;$$

$$\Delta G = -4FE_{cell}$$
$$= 2\Delta_f G^0(NiO) - \Delta_f G(O_2)$$
$$= 2\Delta_f G^0(NiO) - RT\ln p_{O_2}$$
$$RT\ln p_{O_2} = 2\Delta_f G^0(NiO) + 4FE_{cell}.$$

At a temperature of 1100 K, the electrolyte is a good conductor and the equilibrium is established in a reasonable time. At 1100 K, $\Delta_f G^0(NiO) = -139.893$ kJ mol^{-1} (calculated with the HSC Chemistry 6.1). If air at about 1 bar is passed through the cell and equilibrium is attained, E_{cell} can be measured. From the last equation, it follows that at 1100 K:

$$\ln p_{O_2} = -30.526 + 42.20\, E_{cell}.$$

If we express E_{cell} in mV (Figure 9.4), then it is possible to decide whether the sports journalists are right or wrong. The %(v/v) oxygen in the air can be calculated from the quotient of p_{O_2} and the actual pressure of the air in the cell.

Example 9.4

The zirconium dioxide, or zirconia, lambda sensor is based on a solid-state electrochemical fuel cell called the Nernst cell. Its two electrodes provide an output voltage corresponding to the quantity of oxygen in the exhaust relative to that in the atmosphere.

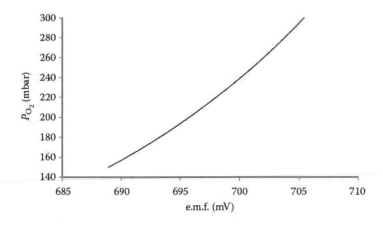

FIGURE 9.4 Oxygen pressure as a function of the e.m.f. for the galvanic cell of the type Ni,NiO│ZrO$_2$(+CaO)│O$_2$.

The sensing element is usually made with a zirconium ceramic bulb coated on both sides with a thin layer of platinum (Figure 9.5) and comes in both heated and unheated forms. The most common application is to measure the performance of internal combustion engines in automobiles and other vehicles.

Lambda probes are used to reduce vehicle emissions by ensuring that engines burn their fuel efficiently and cleanly. The lambda sensor is inserted into the exhaust system of a gasoline engine to measure the volume of oxygen (O$_2$) remaining in the exhaust gas to allow an electronic system [engine control unit (ECU)] to control the efficiency of the combustion process in the engine. The ECU uses feedback from the sensor to adjust the fuel/air mixture.

In most modern automobiles, these sensors are attached to the engine's exhaust manifold to determine whether the mixture of air and gasoline going into the

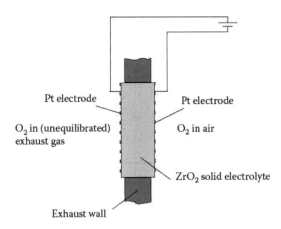

FIGURE 9.5 A sketch of lambda probe for control of a car catalyst.

engine is rich (too much fuel) or lean (too little fuel). An output voltage of 0.2 V (200 mV) DC represents a "lean mixture" of fuel and oxygen, where the amount of oxygen entering the cylinder is sufficient to fully oxidize the carbon monoxide (CO), produced in burning the air and fuel, into carbon dioxide (CO_2). An output voltage of 0.8 V (800 mV) DC represents a "rich mixture," one that is high in unburned fuel and low in remaining oxygen. The ideal setpoint is approximately 0.45 V (450 mV) DC. This is where the quantities of air and fuel are in the optimum ratio such that the exhaust output contains minimal carbon monoxide.

By measuring the amount of oxygen in the remaining exhaust gases, and by knowing the volume and temperature of the air entering the cylinders, among other things, an ECU can use lookup tables to determine the amount of fuel required to burn at the stoichiometric point (14.6:1 air/fuel by mass) to ensure complete combustion.

Failure of these sensors, either through normal aging or the use of leaded fuels, can lead to damage of an automobile's catalytic converter and expensive repairs.

The cell notation is:

$$\ominus O_2(\text{exhaust gas})|ZrO_2|O_2(\text{air})\oplus$$

The electrode reactions are:

$$\ominus: \quad 2O^{2-} \rightarrow O_2(\text{exhaust gas}) + 4e$$

$$\oplus: \quad O_2(\text{air}) + 4e \rightarrow 2O^{2-}.$$

The net chemical reaction of the cell for the passage of four faradays is:

$$O_2(\text{air}) \rightarrow O_2(\text{exhaust gas})$$

$$E_{cell} = E_{right} - E_{left}$$

$$E_{right} = E^0_{O_2} + \frac{RT}{4F} \ln \frac{a_{O_2(\text{air})}}{\left(a_{O^{2-}(ZrO_2)}\right)^2}$$

$$E_{left} = E^0_{O_2} + \frac{RT}{4F} \ln \frac{a_{O_2(\text{exhaust gas})}}{\left(a_{O^{2-}(ZrO_2)}\right)^2}$$

$$E_{cell} = E_{right} - E_{left} = \frac{RT}{4F} \ln \frac{a_{O_2(\text{air})}}{a_{O_2(\text{exhaust gas})}} = \frac{RT}{4F} \ln \frac{p_{O_2(\text{air})}}{p_{O_2(\text{exhaust gas})}}.$$

9.4.1 IONIC ACTIVITIES

Complete cell equations never contain one ion at a time, so individual ionic activities cannot be determined. For example, Equation 9.7 shows that $a_{Fe^{3+}}$ occurs together with a_{Cr} and $a_{Fe^{2+}}$, and this situation is typical of others.

Let us focus attention on the simple 1:1 electrolyte H^+Cl^-. Its activity, a, as a compound, may be found from colligative properties, but the activities of the ions, a_+ and a_-, are not determinable separately. These ionic activities must first be carefully defined in terms of a reference state with unit fugacity; as before, $a_+ = f_+/f_+^0$. The reference states are defined such that, for the equation:

$$HCl \rightleftharpoons H^+ + Cl^-$$

$$(a) \quad (a_+) \ (a_-),$$

the free energy change is zero, and the equilibrium constant $K = 1$. Therefore $a_+ a_-/a = 1$ and $a = a_+ a_-$. Because a_+ and a_- cannot be separately determined, we define the mean ionic activity as $a_\pm = (a_+ a_-)^{1/2}$. Also,

$$\gamma_\pm = (\gamma_+ \gamma_-)^{1/2}$$

and

$$m_\pm = (m_+ m_-)^{1/2}. \tag{9.8}$$

We now have activity, molality, and activity coefficient defined in terms of the geometric mean of the ionic values with, as usual:

$$a_\pm = (\gamma_\pm m_\pm).$$

(This applies to 1:1 electrolytes only, but similar arguments apply to polyfunctional ions.) Note that if the electrolyte is substantially dissociated, then $m_+ = m_- = m$, and $m_\pm = m$.

9.4.2 Analysis of e.m.f. Data to Find E^0

Standard electrode potentials are based on electrolytes of unit activity, with a "hypothetical ideal molality of one." Raw laboratory data arrive as simple e.m.f. values versus molality. How do we find E^0?

Some data of this kind for the cell:

$$Pt, H_2 \text{ (1 bar)}|H^+Cl^-(m)|Hg_2Cl_2, Hg$$

are given in Table 9.2. For this cell, the reaction and Equation 9.5 inform us that:

$$\tfrac{1}{2}H_2(g) + \tfrac{1}{2}Hg_2Cl_2(s) \rightarrow H^+ + Cl^- + Hg(l)$$

$$\text{activity} = \quad 1 \qquad\qquad 1 \qquad\quad a_+ \ \ a_- \quad 1$$

TABLE 9.2
e.m.f. Data at Various Molalities for the Cell Composed of Hydrogen and Calomel Half-Cells (25°C)

Molality, m (mol kg⁻¹ water)	E (V)	$m^{1/2}$	E + 0.1183 log m
0.07508	0.4119	0.2740	0.2789
0.03769	0.4452	0.1941	0.2768
0.01887	0.4787	0.1374	0.2747
0.00504	0.5437	0.0710	0.2719
0.00200	0.5900	0.0447	0.2707

and

$$E = E^0 - \frac{RT}{F} \ln a_+ a_-.$$

We have already decided that $a_+ a_- = a_\pm^2$, and $a_\pm = \gamma_\pm m_\pm = \gamma_\pm m$, so

$$E = E^0 - \frac{2RT}{F} \ln \gamma_\pm m = E^0 - 2(0.05916) \log \gamma_\pm m.$$

Thus,

$$E^0 = \underbrace{E + 0.1183 \log m}_{\substack{\text{Approximately} \\ \text{constant}}} + \underbrace{0.1183 \log \gamma_\pm}_{\substack{\text{Correction term,} \\ \text{approaching zero at} \\ \text{low concentration}}}. \qquad (9.9)$$

It seems that E^0 is approximately equal to the expression $(E + 0.1183 \log m)$, and that it approaches it more and more closely as m tends to zero, and $\gamma_\pm$ tends to one. This term is presented in column 4 of Table 9.2. In the approach to electrolytic solutions pioneered by Debye and Hückel, it was shown that $\log \gamma_\pm$ varied linearly with $m^{1/2}$, for weak solutions. This means that $(E + 0.1183 \log m)$ should show a similar variation; this suspicion is confirmed by Figure 9.6, which shows that $(E + 0.1183 \log m)$ varies approximately linearly with $m^{1/2}$. The resulting value of E^0, obtained by extrapolation to ideal conditions, is 0.2685 V and is in good agreement with other determinations. Once E^0 is known, it is a simple matter to calculate the activity coefficient, $\gamma_\pm$, at a variety of concentrations, from Equation 9.9. At a molality of 0.07508, $\gamma_\pm$ is 0.823. This should be checked and other data calculated.

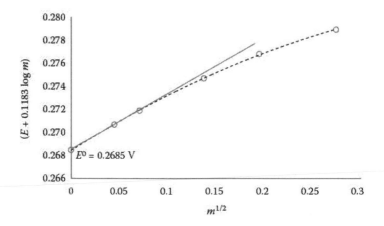

FIGURE 9.6 Plot to determine E^0 accurately for the hydrogen/hydrochloric acid calomel cell. The term $(E + 0.1183 \log m)$ approximates E^0 more closely as concentrations approach zero.

9.5 VARIATION OF E.M.F. WITH TEMPERATURE

One of the most rewarding aspects of electrochemistry is that ΔS values spring directly from studies on the temperature variation of e.m.f. Equation 10.1, to be derived in Chapter 10, describes the temperature variation of ΔG:

$$\left(\frac{d(\Delta G)}{dT}\right)_P = -\Delta S. \tag{10.1}$$

If we substitute for ΔG from Equation 9.2, which is $\Delta G = -nFE$, we have:

$$\left(\frac{d(-nFE)}{dT}\right)_P = -\Delta S,$$

whence,

$$nF\left(\frac{dE}{dT}\right)_P = \Delta S, \tag{9.10}$$

and also

$$nF\left(\frac{dE^0}{dT}\right)_P = \Delta S^0. \tag{9.11}$$

This provides a basically simple means of ΔS determination.

Example 9.5

Worrell [4] at the University of California has studied cells with solid electrolytes that are O^{2-} ionic conductors. Niobium oxides were studied by use of the cell:

$$\ominus Nb, NbO|ThO_2 (Y_2O_3 2\%)|NbO, NbO_2 \oplus.$$

The electrode reactions are:

$$Nb + O^{2-} \rightarrow NbO + 2e \text{ and } NbO_2 + 2e \rightarrow NbO + O^{2-},$$

which give overall $Nb + NbO_2 \rightleftharpoons 2\ NbO$.

Values of e.m.f. at various temperatures are given in Table 9.3. Figure 9.7 shows a graph of E versus T, from which a value for dE/dT of -8.5×10^{-5} V K^{-1} is obtained. Using Equation 9.10, this gives:

$$\Delta S = 2(96,485)(-8.5 \times 10^{-5}) = -16.4 \text{ J mol}^{-1} \text{ K}^{-1}.$$

In addition, values of ΔG are available at any temperature, calculated as $-nFE$. At 1200 K:

$$\Delta G = -2(96,485)(0.201) = -38.79 \text{ kJ mol}^{-1}.$$

Then:

$$\Delta H = \Delta G + T\Delta S = -38.790 + 1200(-16.4)10^{-3} = -58.47 \text{ kJ mol}^{-1}.$$

Finally, all the available thermodynamic information may be summarized in terms of the equation:

$$\Delta G = \Delta H - T\Delta S$$

$$= -58,740 + 16.4T \text{ (J mol}^{-1}).$$

TABLE 9.3
Values of e.m.f. from 1050 to 1300 K for a Solid Electrolyte Cell

E (mV)	T (K)
213	1054
212	1086
209	1106
201.5	1160
202	1174
200	1214
198	1244
194	1283

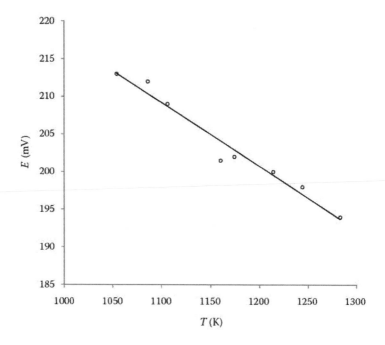

FIGURE 9.7   Graph of e.m.f. against temperature for the "Worrell" cell, from 1054 to 1283 K. The cell is solid throughout.

Example 9.6

Because of uncertainties in published measurements, Wijbenga [5] at The Netherlands Energy Research Foundation at Petten determined the Gibbs energies of formation of URh_3 by measuring the e.m.f. of the cell:

$$Rh, URh_3 \cdot UF_4|CaF_2|NiF_2, Ni.$$

The electrode reactions are:

$$URh_3 + 4F^- \rightarrow UF_4 + 3Rh + 4e \text{ and } 2NiF_2 + 4e \rightarrow 2Ni + 4F^-,$$

which give overall:

$$2NiF_2 + URh_3 \rightarrow 2Ni + UF_4 + 3Rh.$$

Values of e.m.f. and calculated standard Gibbs energies are given in Table 9.4. For the cell reaction:

$$\Delta G = -4(96,485)E_{cell} = \Delta_f G^0 (UF_4) - 2\Delta_f G^0 (NiF_2) - \Delta_f G^0 (URh_3).$$

TABLE 9.4
Values of e.m.f. and $\Delta_f G^0$ (URh$_3$) from 948 to 1114 K for the Galvanic Cell of the Type: Rh, URh$_3$, UF$_4$|CaF$_2$|NiF$_2$, Ni

T (K)	E (mV)	$\Delta_f G^0$ (URh$_3$) (kJ mol^{-1})
948.6	815.7	−307.35
966.8	816.5	−307.40
968.3	816.6	−307.39
982.9	818.0	−307.15
991.1	818.2	−307.24
992.5	817.5	−307.54
997.3	818.6	−307.21
999.2	819.2	−307.02
1000.8	818.4	−307.36
1006.7	820.0	−306.86
1008.2	820.3	−306.78
1024.5	821.5	−306.65
1027.4	821.5	−306.70
1031.5	821.0	−306.98
1036.4	820.3	−307.34
1038.6	820.3	−307.39
1044.2	824.3	−305.96
1044.8	821.8	−306.94
1065.5	825.4	−305.97
1067.7	823.0	−306.93
1089.0	825.6	−306.36
1113.2	828.3	−305.81

The standard molar Gibbs energy of formation of URh$_3$ can be calculated from:

$$\Delta_f G^0 \text{ (URh}_3) = \Delta_f G^0 \text{ (UF}_4) - 2\Delta_f G^0 \text{ (NiF}_2) + 4(96{,}485)E_{cell}.$$

The Gibbs energies of formation of NiF$_2$ and UF$_4$ as functions of temperature are well established:

$$\Delta_f G^0(\text{NiF}_2) = -653.50 + 0.15129T \text{ (kJ mol}^{-1}; 800-1100 \text{ K).}$$

$$\Delta_f G^0(\text{UF}_2) = -1910.0 + 0.2823T \text{ (kJ mol}^{-1}; 800-1100 \text{ K).}$$

If we substitute the last two equations in the equation of $\Delta_f G^0$ (URh$_3$), we obtain:

$$\Delta_f G^0 \text{ (URh}_3) = -602.96 - 0.020233T + 385.94E_{cell} \text{ (kJ mol}^{-1})$$

With this equation the standard molar Gibbs energy of formation of URh_3 can be calculated for each temperature, and is given in Table 9.4. With the linear regression function on a pocket calculator or with Excel, the following equation is derived for $\Delta_f G^0$ (URh_3) as function of temperature:

$$\Delta_f G^0(URh_3) = -316.37 + 0.009257T \text{ (kJ mol}^{-1}; 950 - 1115 \text{ K)}.$$

(This should be checked by the student!)

Over the temperature range 950–1115 K the average values of $\Delta_f H^0$ (URh_3) and $\Delta_f S^0$ (URh_3) are −316.37 kJ mol^{-1} and −9.257 J mol^{-1} K^{-1}, respectively. If C_p data are known there can be calculated using Kirchhoff's equation (Equation 3.4) that $\Delta_f H^0$ (URh_3, 298.15 K) = −301.16 kJ mol^{-1}.

REFERENCES

1. Moore, W. J. 1963. *Physical Chemistry*. 4th ed. London: Longmans.
2. Potter, E. C. 1961. *Electrochemistry*. London: Cleaver-Hume.
3. Latimer, W. 1952. *Oxidation Potentials*. Englewood Cliffs, NJ: Prentice-Hall.
4. Worrell, W. L. 1966. Thermodynamics of nuclear materials. *Symposium on Thermodynamics*, vol. I, pp. 131–143. Vienna: International Atomic Energy Agency.
5. Wijbenga, G. J., and E. H. P. Cordfunke. 1982. Determination of standard Gibbs energies of formation of URh_3 and URu_3 by solid-state e.f.f. measurements. *Chem Thermodyn* 14:409–417.
6. King, J. P., and J. W. Cobble. 1957. Thermodynamic properties of technetium and rhenium compounds. *J Am Chem Soc* 79:1559–1563.
7. Cordfunke, E. H. P., and R. J. M. Konings. 1988. The enthalpy of formation of RuO_2. *Thermochim Acta* 129:63–69.

PROBLEMS

9.1 An iron penknife blade may be copper-plated in $CuSO_4$ solution by displacement of Cu^{2+} ions with Fe^{2+} ions. Assuming that enough penknife blades are available, how low can the Cu^{2+} activity be taken at 25°C if the initial activity is one? (Hint: Start your calculations by setting $a_{Cu^{2+}} = \alpha$ at equilibrium.)

9.2. The compound Ni_2O_3 is a strong oxidizing agent, and is used in the Edison storage cell, which is represented as:

$$\ominus Fe, FeO|KOH \text{ (aq)}|Ni_2O_3, NiO\oplus.$$

The cell reaction is:

$$Fe + Ni_2O_3 \rightarrow FeO + 2NiO.$$

(a) Write down the half-cell reactions.
(b) If the cell e.m.f. is +1.27 volts, what is the free energy of formation of Ni_2O_3? (Data available from Appendix III.)

(c) What effect does KOH activity have on the equilibrium e.m.f.? On the current?

9.3 The following reactions have been used for fuel cells. Write down a cell diagram and electrode reactions for each:

$$2CuBr_2 \rightleftharpoons 2CuBr + Br_2$$

$$PbI_2 \rightleftharpoons Pb + I_2$$

$$2LiH \rightleftharpoons 2Li + H_2.$$

If $\Delta_f G^0 (LiH) = -68.63 \text{ kJ mol}^{-1}$, what equilibrium potential is expected for the third cell?

9.4 Silver ions oxidize ferrous ions according to the equation:

$$Ag^+(aq) + Fe^{2+}(aq) \rightarrow Ag(s) + Fe^{3+}(aq).$$

The equilibrium constant at 298 K is $K = 7.810$. This reaction may also be studied as a cell, such as:

$$Pt|Fe^{2+},Fe^{3+}|NH_4NO_3 \text{ bridge}|Ag^+NO_3^-|Ag.$$

Determine:
(a) the half-cell reactions
(b) the polarity
(c) the expected e.m.f. of the cell with unit activity electrolytes

9.5 The following cell is studied experimentally at 298 K:

$$Pt, H_2(1 \text{ bar})|H^+Cl^-(a = 1)|H_2(0.2 \text{ bar}), Pt$$

What is the cell reaction? What polarity do you suggest? Assuming that $a_{H_2} = P_{H_2}$, calculate the e.m.f.

9.6 The glass electrode is sensitive to the activity of hydrogen ions. If used in conjunction with a calomel reference electrode and a KCl salt bridge, the following equations apply:

$$E = E^0 + \frac{RT}{F} \ln a_{H^+}$$

$$E = E^0 + \frac{2.303RT}{F} pH.$$

And where pH is defined here as being equal to $-\log a_{H^+}$, the following data obtained at 25°C from such a cell enable one to derive a value for F, and Faraday constant, graphically, with the linear regression function on your pocket calculator or with Excel on your PC. Express your answer in coulomb mol^{-1}.

pH	4.01	6.49	6.99	9.15
E (V)	−0.1245	−0.271	−0.3015	−0.430

9.7   Write the electrode reactions, and the total cell reaction for

$$\text{Pt, H}_2(\text{1 atm})|\text{HCl in MeOH}|\text{AgCl, Ag.}$$

At 25°C, data for this cell are:

m (mol kg^{-1} MeOH)	E (V)
2.110×10^{-3}	0.3042
4.265×10^{-3}	0.2713
7.749×10^{-3}	0.2446
1.353×10^{-2}	0.2208

From an appropriate plot of these data, calculate E^0, the standard electrode potential of the silver-silver chloride electrode in methanol solvent.

9.8   King and Cobble [6] found that, for the cell in which the reaction:

$$\tfrac{1}{2}\text{H}_2(\text{g}) + \text{HReO}_4(\text{aq}) \rightarrow \text{ReO}_3(\text{s}) + \text{H}_2\text{O(l)}$$

takes place, $dE^0/dT = -1.21 \times 10^{-3}$ V K^{-1}. If $S^0(\text{HReO}_4(\text{aq})) = 202.1$ J mol^{-1} K^{-1}, what is the molar entropy of ReO$_3$(s)?

9.9   A student study of the cell Pt, H$_2$(1 bar)|H$^+$Cl$^-$ ($c = 1$)|AgCl, Ag$^+$ produced the following data of e.m.f. versus temperature:

T (°C)	e.m.f. (mV)	T (°C)	e.m.f. (mV)
26	206	40	197
28	205	42	196
30	204	44	194
32	202	46	193
34	201	48	191
36	199.5	50	190
38	198	52	188.5

Determine ΔS^0 for the cell reaction by graphical means or with Excel on your PC and compare with ΔS^0_{298} calculated from Appendix III.

9.10 As part of a systematic investigation of the thermochemical properties of compounds of fission products, Cordfunke and Konings [7] studied the thermochemical properties of RuO$_2$. Moreover, since ruthenium is formed in high yields during the fission of uranium in a nuclear reactor it also plays an important role in analysis of accidents in nuclear reactors. The Gibbs energy of formation of RuO$_2$(s) has been determined from 962 to 1070 K by the e.m.f. technique using ZrO$_2$(+CaO) as the solid electrolyte in the reversible cell:

$$\text{Ru, RuO}_2|\text{ZrO}_2(+\text{CaO})|\text{O}_2(p \approx 1 \text{ bar}).$$

The net chemical reaction of the cell for passage of 4 faradays is:

$$Ru(s) + O_2(g) \rightarrow RuO_2(s).$$

One of the measured series is given below:

T (K)	E (mV)	p_{O_2} (Pa)
986.8	349.9	98,712
1011.4	338.8	99,645
1036.7	328.1	101,245
1006.5	341.5	101,925

Determine:
(a) the half-cell reactions
(b) $\Delta_f G^0$ ($RuO_2(s)$) for each temperature
(c) from the $\Delta_f G^0$ (RuO_2) data by linear regression $\Delta_f G^0$ (RuO_2) as a function of temperature
(d) the average $\Delta_f H^0$ (RuO_2) and $\Delta_r S^0$ in the temperature range 986–1007 K

9.11 The cell represented by Pb, $PbI_2|K^+I^-$ solution|AgI, Ag has an e.m.f. of +0.2078 V at 25°C, and $(dE/dT)_p = -1.88 \times 10^{-4}$ V K^{-1}. Write down the cell reaction and calculate ΔG, ΔS, and ΔH. It will be found that all components appearing in K are solids; why therefore is not $K = 1$, $\Delta G = 0$?

9.12 At 25°C, the e.m.f. of the cell:

$$Pb|Pb^{2+}|5\% \text{ Pb amalgam}|Hg$$

is 5.7×10^{-3} V and its temperature coefficient is 1.6×10^{-4} V K^{-1}. Calculate:
(a) the entropy change when current is drawn reversibly from the cell
(b) the associated enthalpy change

9.13 The Weston standard cell provides 1.0186 V at 20°C, and:

$$(dE/dT)_p = -4.06 \times 10^{-5} \text{ V K}^{-1}.$$

Calculate ΔG_{293} for the process:

$$\tfrac{1}{2}Cd + \tfrac{1}{2}Hg_2SO_4 \rightarrow \tfrac{1}{2}CdSO_4(s) + Hg,$$

which is the spontaneously occurring change. Find also ΔS. Why is the Weston cell so well suited to be a standard?

9.14 The sensors of an oxygen sensor (lambda probe) from Example 9.4 only work effectively when heated to approximately 800°C, so most newer lambda probes have heating elements encased in the ceramic to bring the ceramic tip up to temperature quickly when the exhaust is cold.

At 800°C, calculate the amount of oxygen in the exhaust gas of a car relative to the amount of oxygen in air for:
(a) a "lean" mixture of fuel and oxygen
(b) a "rich" mixture of fuel and oxygen
(c) a mixture of fuel and oxygen at "stoichiometric point"

9.15 To determine the solubility product of AgCl at 25°C, the following cell is studied:

$$Ag(s)|AgCl(aq)|AgCl(s)|Ag(s).$$

The cell reaction is:

$$AgCl(s) \rightleftarrows Ag^+(aq) + Cl^-(aq).$$

Determine:
(a) the half-cell reactions
(b) the expected e.m.f. of the cell with unit activity electrolytes
(c) the solubility product $K_{s,AgCl} = a_{Ag^+} \cdot a_{Cl^-}$ at 25°C from the calculated e.m.f.
(d) from $\Delta_f G^0$ values from Appendix III, the solubility product of AgCl at 25°C

10 Free Energy and Industrial Processes

At the start of this chapter, we shall attempt to describe accurately the changes of ΔG with temperature. This will require careful computation of ΔC_P. Then, using free energy data, we shall discuss the outcome of four chemical processes, all operated on an industrial scale. These are the direct synthesis of ethanol from ethylene, which has now largely superseded the fermentation processes; the Pidgeon process for the production of magnesium; the process used for titanium manufacture; and silicon metal production in an electric arc furnace. Ellingham diagrams will be introduced and applied, and a practical assessment made of the importance of activity coefficients.

10.1 FREE ENERGIES AS A FUNCTION OF TEMPERATURE

We have often used the equation defining free energy changes as:

$$\Delta G = \Delta H - T\Delta S. \tag{6.3}$$

As a rule, when calculating ΔG at temperatures other than 298 K, we have made the assumption that ΔH and ΔS are independent of temperature. That is, we have calculated free energy changes at any temperature T as:

$$\Delta G_T^0 = \Delta H_{298}^0 - T\Delta S_{298}^0.$$

and although this is adequate for small temperature ranges, significant errors are introduced when, for example, high-temperature metallurgical processes are described. Although individual variations in ΔH and ΔS have been accounted for in theory (e.g., Section 8.1), no general expression for ΔG at any temperature has been attempted. We shall now remedy this.

10.1.1 THE GIBBS-HELMHOLTZ EQUATION

We start with Equation 7.1, which asserts that:

$$dG = VdP - SdT. \tag{7.1}$$

In constant pressure situations, $dP = 0$, and we may write

$$\left(\frac{dG}{dT}\right)_P = -S.$$

161

Alternatively, for a reaction, we could modify this to read:

$$\left(\frac{d(\Delta G)}{dT} \right)_P = -\Delta S. \tag{10.1}$$

This expression has great usefulness and has already been exploited in the previous chapter. On introducing this into Equation 6.3, we have one form of the *Gibbs-Helmholtz equation*:

$$\Delta G = \Delta H + T \left(\frac{d(\Delta G)}{dT} \right)_P. \tag{10.2}$$

In the discussion that follows, we shall find a preferable alternative expression.

Let us now differentiate $\Delta G/T$, with respect to temperature. From the usual rule for differentiating a quotient, we find that:

$$\frac{d(\Delta G/T)}{dT} = \frac{T(d(\Delta G)/dT) - \Delta G}{T^2}.$$

If we now substitute for ΔG from Equation 10.2, we have:

$$\frac{d(\Delta G/T)}{dT} = -\frac{\Delta H}{T^2}. \tag{10.3}$$

This is merely an alternative expression for Equation 10.2. However, as they stand, neither of these alternatives can help us to find high-temperature values of ΔG accurately. Clearly, we must integrate one of these expressions.

10.1.2 THE INTEGRATED FORM OF THE GIBBS-HELMHOLTZ EQUATION

To integrate Equation 10.3, we start by writing it as:

$$\int d(\Delta G/T) = -\int \frac{\Delta H}{T^2} dT. \tag{10.4}$$

We are now obliged to express ΔH as a function of temperature. From Section 3.2, we recall that $(dH/dT)_P = C_P$, and that heat capacities were expressed as simple polynomials in T, of the form $C_P = a + bT + cT^2 + dT^3$. For reactions, we write:

$$\left(\frac{d\Delta H}{dT} \right)_P = \Delta C_P = \Delta a + \Delta bT + \Delta cT^2 + \Delta dT^3 \tag{10.5}$$

where Δ has the usual meaning of "products minus reactants." On integrating:

$$\int d(\Delta H) = \int \left(\Delta a + \Delta bT + \Delta cT^2 + \Delta dT^3 \right) dT.$$

$$\Delta H_2 - \Delta H_1 = \left[\Delta aT + \Delta bT^2/2 + \Delta cT^3/3 + \Delta dT^4/4 \right]_{T_1}^{T_2}.$$

In general,

$$\Delta H_T = \Delta H_0 + \Delta aT + \Delta bT^2/2 + \Delta cT^3/3 + \Delta dT^4/4 \qquad (10.6)$$

where ΔH_0 is the value of ΔH_T when $T = 0$ K. One must be careful to notice the difference between the zero as superscript, meaning standard state, and as subscript, meaning zero temperature. This expression for ΔH_T may now be inserted into Equation 10.4 to give, on integration:

$$\Delta G_T = \Delta H_0 - \Delta aT \ln T - \Delta bT^2/2 - \Delta cT^3/6 - \Delta dT^4/12 + IT. \qquad (10.7)$$

The term I is an integration constant yet to be determined. It should be stressed that this impressive equation can be taken as seriously, or as lightly, as the circumstances demand. In some cases, the accuracy will only justify an averaged value of Δa, later terms being neglected. However, even when simplified to this extent, and especially when taken rigorously, it can involve a great deal of work. Suppose we know ΔG_{298}, ΔH_{298}, and the expressions for C_p for each reactant and product. We need to know ΔH_0, and I before proceeding with the ΔG_T evaluation. The scheme of calculation would be somewhat as follows:

(a) Determine Δa, Δb, and so on from the C_p expressions.
(b) Use Equation 10.6, the known value for ΔH_{298}, and $T = 298$ K, and calculate ΔH_0.
(c) Use Equation 10.7, the value of ΔH_{298}, and, ΔH_0 with $T = 298$ K, and find I.
(d) Calculate ΔG at any desired temperature.

10.1.3 TABULATED FORMS OF FREE ENERGY

The persistent optimist will surely say, "There must be an easier way." There are three developments that we might examine before moving to some practical applications. The first possibility is to tabulate ΔG^0 over a range of temperatures, say every 200°C, for all formation reactions and then find the desired ΔG^0 values by interpolation. This is a good idea in theory, but in practice is simply not accurate enough for detailed calculations of equilibrium constants. However, it is a very useful aid in the preliminary assessment of untried processes. We shall say more about this kind of approach later, when discussing Ellingham diagrams. The second possibility is to find a function that involves ΔG^0 but does not change with temperature as violently as does ΔG^0 itself. It has been found that the free energy function (f.e.f.), $-(G^0 - H_0^0)/T$, which is assignable to individual compounds and elements rather than reactions, does just this. It has the additional advantage that it may be calculated from spectroscopic data, which are inherently reliable and accurate. The reader is referred to Lewis and Randall's [1] book for tabulations of f.e.f. values and further discussion of them. The third possibility is the use of modern computer programs. Calculations using thermochemical databases can often be carried out in a matter of minutes at the terminal. The already mentioned computer program HSC Chemistry can calculate ΔG_T^0 and other thermochemical functions for chemical reactions in a few seconds. All examples and problems in this book have been checked by this program. In the next chapter, we will take a closer look at computer-assisted thermochemistry.

10.2   THE SYNTHESIS OF ETHANOL

Most people think that industrial ethanol is produced from the fermentation of carbohydrates such as molasses and grains. But almost all industrially produced ethanol today is synthesized from the hydration of ethylene. Ethanol made from ethylene is purer and cheaper.

Of the 760×10^6 L of industrial ethanol (excluding fuels) produced in the United States in 1981, less than 2% was made by fermentation. Carbohydrate feedstock sources normally used for fermentation are prone to continually changing costs and cause major distortions to the price of the end product. The fermentation route was attractive while cheap Cuban molasses were available, and fermentation may be attractive again if there are petroleum shortages in future. The process is based on the direct hydration of ethylene:

$$C_2H_4 + H_2O \rightarrow C_2H_5OH$$

at 6.8 MPa pressure, and nearly 600 K, over a phosphoric acid catalyst. We shall discuss certain aspects of this reaction to exemplify detailed ΔG calculations, and also the application of activity coefficients (Chapter 7) to high-pressure processes.

10.2.1   Equilibrium Calculations

We shall now follow the recipe given at the end of Section 10.1.2. First, we find that the C_P functions are (units are in J mol^{-1} K^{-1}, data from Table 3.3):

$$C_P(C_2H_5OH) = 19.9 + 20.95 \times 10^{-2}T - 10.37 \times 10^{-5}T^2 + 20.04 \times 10^{-9}T^3$$

$$C_P(C_2H_4) = 3.95 + 15.63 \times 10^{-2}T - 8.339 \times 10^{-5}T^2 + 17.66 \times 10^{-9}T^3$$

$$C_P(H_2O) = 29.16 + 1.449 \times 10^{-2}T - 0.2022 \times 10^{-5}T^2,$$

and so

$$\Delta C_P = C_P(C_2H_5OH) - C_P(C_2H_4) - C_P(H_2O)$$

$$= -31.21 + 3.871 \times 10^{-2}T - 1.8359 \times 10^{-5}T^2 + 2.38 \times 10^{-9}T^3.$$

That is, in Equation 10.6, $\Delta a = -13.21$, and so on.

Second, we need to know ΔH^0_{298} from enthalpy of formation data. This is found to be -45.41 kJ mol^{-1}, and so we can now use Equation 10.6, with $T = 298$ K. Thus,

$$\Delta H^0_0 = \Delta H^0_{298} - \Delta a(298) - \Delta b(298^2)/2 - \Delta c(298^3)/3 - \Delta d(298^4)/4$$

$$= -43.025 \text{ kJ mol}^{-1}$$

Third, we can use Equation 10.7 to find I, if we have ΔH^0_0 and ΔH^0_{298}. From the free energies of formation given in Appendix III, $\Delta G^0_{298} = -7.53$ kJ mol^{-1}. As a result:

$$I = \Delta G^0_{298}/298 - \Delta H^0_0/298 + \Delta a \ln 298 + \Delta b(298)/2$$

$$+ \Delta c(298^2)/6 + \Delta d(298^3)/12$$

$$= +47.796 \text{ J mol}^{-1} \text{ K}^{-1}.$$

Finally, our ΔG^0_T Equation 10.7 now reads:

$$\Delta G^0_T = -43025 - \Delta aT \ln T - \Delta bT^2/2 - \Delta cT^3/6 - \Delta dT^4/12 + 47.796T.$$

Taking all the plant parameters into account, the Shell Development Company [2] recommends operation at 572 K. In this case, we find:

$$\Delta G^0_{572} = +26.972 \text{ kJ mol}^{-1}$$

and

$$K = 3.44 \times 10^{-3}.$$

We must temper our initial disappointment at the small equilibrium constant by recalling that industrial processes often involve recycling of unreacted starting material. This depends on effective removal of what product there is, before the unreacted ethylene is put through the catalyst bed once again. In this case, washing of product mixture with liquid water is all that is required to remove the ethanol, and so commercial operation is practicable after all.

Before making accurate calculations of the equilibrium mixture, it is interesting to compare the value for ΔH^0_{572} obtained above with the one calculated from the simple $\Delta G^0 = \Delta H^0 - T\Delta S^0$ equation, and to compare them both with the definitive value obtained from experimental equilibrium measurements.

First, using

$$\Delta G_T^0 = \Delta H_{298}^0 - T\Delta S_{298}^0$$

we find that

$$\Delta G_{572}^0 = +26.958 \text{ kJ mol}^{-1}$$

and

$$K = 3.45 \times 10^{-3}.$$

The experimental value [3] of K is 3.14×10^{-3}, giving $\Delta G_{572}^0 = +27.417$ kJ mol^{-1}. We shall adopt this value for future discussion. It will be noticed that although both calculated values of K are slightly higher than the experimental value, the more carefully calculated figure is at least the nearer of the two. All three values are close together, and we must conclude that ΔC_p is sufficiently small to be unimportant in this case. However, only by doing the calculation can we be sure of our derived information.

It is interesting to mention that with the computer program HSC Chemistry, we find in a few seconds a value $\Delta G_{572}^0 = 27.707$ kJ mol^{-1}, and $K = 2.95 \times 10^{-3}$. This is very close to the experimental value.

10.2.2 Use of Activity Coefficients

In Section 7.7, we discussed the derivation and use of the equation $\Delta G^0 = -RT \ln K$. In general, for each gas in a gas phase re action, $f = \gamma p$. In the case where pressure is low, the fugacity approaches the pressure, and $\gamma = 1$. However, in the ethanol process, pressures are high, and we must express K as:

$$K = \frac{f_{alc}}{f_e f_w} = \frac{p_{alc}}{p_e p_w} \frac{\gamma_{alc}}{\gamma_e \gamma_w}. \tag{10.8}$$

The subscripts alc, e, and w refer to alcohol, ethylene, and water, respectively. The term in pressures is usually written K_p, and the partial pressures are derived from the total pressure multiplied by mole fraction. The expression in activity coefficients, because of its similarity in form to K_p, is often abbreviated K_γ; it is not, of course, an equilibrium constant. In short, we may write $K = K_p K_\gamma$. The activity coefficients are determined from the generalized chart shown as Figure 7.7, from known or estimated pressures. However, to start with, we know only K and the total pressure, and must estimate starting values of each pressure in order to find values for γ in each case. We can then estimate K_p as K/K_γ, and calculate equilibrium pressures once more. If the new values for γ are appreciably different, then another iteration (calculation) is called for. The approach is similar to that of a golfer approaching his hole; he arrives by successive approximation. Let us see how this works out in practice.

The following parameters are quoted for the ethanol process:

Mole ratio of feedstock: ethylene 10, water vapor 6
Temperature: 572 K
Total pressure: 67.2 bar

The mole fraction and partial pressure situation is summarized as follows:

$$C_2H_4 \quad + \quad H_2O \quad \rightleftarrows \quad C_2H_5OH.$$

Initial moles	10	6	0
Initial partial pressure	(67.2)10/16	(67.2)6/16	0
Moles at equilibrium	$10 - x$	$6 - x$	x
Pressure at equilibrium	$\dfrac{(10-x)67.2}{(16-x)}$	$\dfrac{(6-x)67.2}{(16-x)}$	$\dfrac{67.2x}{(16-x)}$

Therefore,

$$K_P = \frac{x(16-x)}{67.2(10-x)(6-x)},$$

and conversion of ethylene will be $100(x/10)\%$ (for every 10 mol entering the plant, x react).

We begin by assuming $\gamma = 1$ for each gas, and $K\gamma = 1$. Then $K_p = 3.14 \times 10^{-3}$, and we can solve for x. This turns out to be 0.683, and so ethylene conversion is $100(0.683/10)$ or 6.83%. Then, $p_{alc} = 2.96$ bar, $p_e = 40.87$ bar, and $p_w = 23.35$ bar. This now enables us to look up γ values using the data on critical points, and reduced temperatures and pressures shown in Table 10.1: this gives a value of $K_\gamma = 1.067$. The second time around, using our improved value of $K/1.067 = 2.94 \times 10^{-3}$ for K_p, we find new pressures and gammas, which give $K_\gamma = 1.070$. A third and final calculation gives $x = 0.644$, and the results as seen in Table 10.1.

TABLE 10.1
Critical Constants of Ethanol, Ethylene, and Water

	T_c (K)	P_c (atm)	T_r	P_r
Ethanol	516.3	63.1	1.108	0.048
Ethylene	282.9	50.9	2.022	0.792
Water	647.0	217.7	0.884	0.105

Alcohol	$\gamma = 0.99$	$p = 2.79$ bar	$f = 2.76$ bar
Ethylene	$\gamma = 0.99$	$p = 40.95$ bar	$f = 40.53$ bar
Water	$\gamma = 0.935$	$p = 23.46$ bar	$f = 21.93$ bar
	$K_\gamma = 1.070$	$K_p = 2.93 \times 10^{-3}$	$K = 3.14 \times 10^{-3}$

Because the gases do not behave ideally, the ethylene conversion is reduced from 6.83% to 6.44%, which represents a significant lowering in the expected yield.

It is a small change, however, when compared with other reactions; moreover, nonideality may lead to an improvement in K_p, as in the ammonia synthesis. We find that K_p for the reaction:

$$\tfrac{1}{2}N_2 + \tfrac{3}{2}H_2 \rightleftarrows NH_3$$

is 0.0070 at 450°C and 1 atm, but has risen to 0.013 for a total pressure of 600 atm. In this case, therefore, the yield will increase on two counts. First, high pressure helps conversion to low-volume products, and second, the gas imperfections happen to make a further change for the better.

It is interesting that the ethanol plant runs with about 5% yield, which means that the catalyst (phosphoric acid on clay pellets) is not fully effective. The remaining 95% of unreacted ethylene is recycled with an additional 5% of fresh feed. On average, therefore, an ethylene molecule will make 19 trips before reacting successfully on the 20th.

10.3   ELLINGHAM DIAGRAMS

Calculations of ΔG_T^0 using Equation 10.3 are excessively time-consuming, but for high-temperature processes such as occur in metal extraction and nuclear power plants, data must be available over wide ranges of temperature. It is therefore necessary to take rigorous account of the temperature variation of ΔG^0, even though the variation often turns out to be approximately linear. It was Ellingham [4], who suggested that ΔG^0 data for many reactions of one class, say oxidations of metals, should be plotted against T in such a form that it could be seen at a glance if a certain metal would reduce the oxide of another. As Ellingham put it:

> It occurred to the present author that a useful purpose would be served by representing graphically the variation with temperature of the standard free energy of formation of compounds of a particular class (say the oxides) on a single diagram. [4]

It would provide, he said, "what might be described as a ground plan of metallurgical possibilities."

The following discussion of Ellingham diagrams owes much to the excellent monograph of Ives [5], to which further reference is recommended. An example of such a chart is shown in Figure 10.1 for oxide formation. Familiarity will come with practice with many different reactions; for the present, consider the line due to

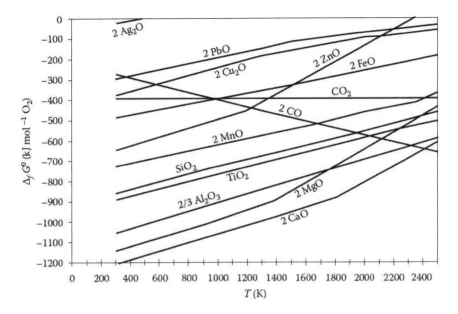

FIGURE 10.1 A simplified Ellingham diagram for oxide formation. It shows $\Delta_f G^0$, based on 1 mol of oxygen, for a number of oxides, as a function of temperature. (Based on results generated by FactSage [6] using the SGTE Pure Substance Database.)

Zn/ZnO. Values of ΔG^0 for all the reactions shown on the diagram involve *1 mol of oxygen*, in the formation of an oxide:

$$\frac{2x}{y}M + O_2 \rightarrow \frac{2}{y}M_xO_y;$$

here

$$2Zn + O_2 \rightarrow 2ZnO.$$

The first point of interest is that the curves consist of a number of apparently straight lines that change slope abruptly at each phase change. The deviation from actual linearity in most cases is within the experimental errors. The full calculations are not wasted effort, however, because simple extrapolations to high temperatures using low-temperature data are not justified. With one extremely important exception, the lines generally have a positive gradient, which means that the oxides become less stable at elevated temperatures. Some oxides, such as Ag_2O have $\Delta G^0 = 0$ at the relatively low temperature of 130°C, with the result that $K = 1$, and the dissociation oxygen pressure is 1 bar.

The second main point concerns the slope of lines. In the case of ZnO formation, the line has a positive slope, which steepens at the boiling point (1180 K). Equation 10.1 relates this slope to ΔS:

$$\left(\frac{d(\Delta G^0)}{dT}\right)_P = -\Delta S^0.$$

The zinc reaction between its melting (693 K) and boiling (1180 K) point:

$$2Zn(l) + O_2(g) \rightarrow 2ZnO(s),$$

involves the consumption of 1 mol of gas, and ΔS^0 is clearly negative (it is -216 J K^{-1} at 1000 K). At the boiling point of zinc, $S^0(Zn)$ increases, and $d(\Delta G)/dT$ becomes much more positive as a result. Vaporization of zinc causes an increase in $(-\Delta S)$, because 3 mol of gas are consumed, and, hence, of slope. Steepening is associated with a greater entropy drop on oxide formation. On the other hand, in the case of Pb/PbO, the oxide boils at 1994 K, and $-\Delta S^0$ for the process:

$$2Pb(l) + O_2(g) \rightarrow 2PbO(g),$$

then actually becomes positive because of the 2 mol of gaseous product. The gradient is negative for a short while until lead boils, when the line sets off uphill once more.

Third, oxidation-reduction processes may be analyzed visually, because all reactions are based on 1 mol of oxygen. At 1000 K, the line for Ti/TiO$_2$ lies well below that for Zn/ZnO. That is, TiO$_2$ has a larger negative value of ΔG^0, and is the more stable of the two oxides. The oxide lower on the chart is more stable. A numerical analysis is simple. For the two oxides, and the reduction, we write:

$$2Zn + O_2 \rightarrow 2ZnO, \qquad \Delta G^0_{1000} = -497 \text{ kJ}, \qquad (I)$$

$$\underline{Ti + O_2 \rightarrow TiO_2 \qquad \Delta G^0_{1000} = -761 \text{ kJ}} \qquad (II)$$

$$2\,ZnO + Ti \rightarrow 2\,Zn + TiO_2 \qquad \Delta G^0_{1000} = -264 \text{ kJ}. \qquad (II - I)$$

Thus, zinc oxide would clearly be (thermodynamically) unstable in contact with hot titanium, although the rate of such corrosion would be extremely slow, due to the extraordinarily tough oxide layer on zinc metal. Moreover, we see that the use of zinc as a reducing agent for TiO$_2$ is quite out of the question, not only at 1000 K, but at any practicable temperature, because the titanium line remains below that of zinc, whatever the temperature.

The only oxide to become consistently more stable as temperature increases is carbon monoxide. On this fact the Iron Age depended; virtually any metal oxide is reducible by carbon at a sufficiently high temperature. Visually, this means that the carbon monoxide line moves underneath those of other oxides, as the temperature increases. Thus manganous oxide (MnO) is stable to hot coke up to 1680 K, at which point the lines for MnO and CO meet, and ΔG^0 for the reaction:

$$MnO + C \rightarrow Mn + CO$$

is zero, since it is equal to $\Delta_f G^0(CO) - \Delta_f G^0(MnO)$. Above this temperature, ΔG^0 becomes negative; at 1750 K,

$$\Delta G^0 = \tfrac{1}{2}\Delta_f G^0(2\,CO) - \tfrac{1}{2}\Delta_f G^0(2\,MnO)$$

$$= \frac{-530.2}{2} - \frac{-506.0}{2}$$

$$= -12.1 \text{ kJ}$$

and the production of liquid manganese metal proceeds. Notice that each value of ΔG^0 refers to 1 mol of O_2. It then becomes possible, although economically unattractive, to reduce other oxides with carbon at yet higher temperatures. Production of silicon becomes feasible at 1800 K, production of titanium at 1860 K, of magnesium at 1910 K, and even of aluminum, zirconium, and calcium between 2350 and 2450 K. Although both titanium and magnesium have been produced by the "carbothermic" method, the cost of such high-temperature operations is very high, and alternative processes are now in use. In particular, electrolytic methods become increasingly attractive in this situation. A free energy change of 600 kJ mol^{-1} O_2 may be overcome by a mere 1.6 V of applied external potential, and so wherever cheap electricity is available, such as in Canada or Ghana (Volta River Project), large aluminum plants may spring up. The equilibrium potential will be simply $E^0 = -\Delta G^0/nF$, although higher voltages than this are invariably used. For an appreciable current, the cell resistance and electrode polarization must be overcome; the resistive heating is often sufficient to keep the electrolyte molten.

Carbon has two oxides of comparable stability, and as a result shows dual behavior. Formation of monoxide, as we have seen, gives a downward sloping line. The formation of carbon dioxide:

$$C + O_2 \rightarrow CO_2$$

has $\Delta S^0 = +2.9$ J K^{-1}, and gives an almost horizontal line, which cuts that of the monoxide at 973 K. Below this temperature formation of CO_2 predominates; above it, CO. At 973 K, they are sharing the honors. The (Boudouard) disproportionate reaction:

$$C + CO_2 \rightarrow 2CO$$

has $\Delta G^0 = \Delta_f G^0(2CO) - \Delta_f G^0(CO_2) = 0$, and $K = 1$. An additional line, representing the oxidation of CO:

$$2CO + O_2 \rightarrow 2CO_2$$

is important during low-temperature reductions, because CO is the gas phase intermediate. The monoxide runs a shuttle service for oxygen between the solid oxide and the solid-coke reducing agent, but is not itself an overall reaction product. We are now able to discuss five applications, two briefly, and three in some detail.

10.3.1 CORROSION PREVENTION

Steam turbines can become more efficient with hotter primary steam. This in turn requires high-temperature furnaces, and in nuclear power stations, fuel elements that can withstand higher temperatures. Liquid sodium has been used in nuclear reactors as a primary, closed-loop, cooling medium, but the possibility of it corroding the stainless steel used to contain the uranium oxide fuel elements has caused concern.

The oxidation of multicomponent alloys is a complex process from thermodynamic and kinetic points of view. To a great extent, the chromium content determines the oxidation behavior of stainless steel [7]. Figure 10.2 shows clearly that at about 1000°C, sodium oxide can oxidize stainless steel. (The dashed line represents liquid sodium above its normal boiling point.) A diagram of this kind shows not only that sodium can cause material transport of oxygen from a point of external contamination, but also that metallic zirconium is able to "hot-trap" the oxygen from the sodium stream.

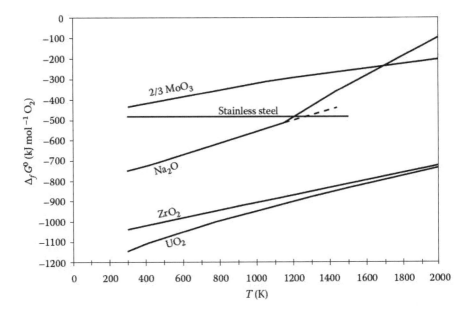

FIGURE 10.2 Ellingham free energy diagram showing possibility of corrosion of stainless steel fuel cans by sodium oxide near 1250 K. Liquid sodium coolant may be kept free of oxygen by zirconium metal, which has a greater affinity for oxygen. (Except for the stainless steel line, all data are based on results generated by FactSage [6] using the SGTE Pure Substance Database.)

10.3.2   ELECTROLYSIS OF ALUMINA

Paul Louis Toussaint Héroult obtained a patent in April 1886, just 7 weeks before the American Charles Hall obtained his, for processes to make aluminum from bauxite. They were independently developed, but almost identical. A molten bath of bauxite, Al_2O_3, in cryolite, $Na_3[AlF_6]$, is electrolyzed with carbon electrodes at about 1240 K. Molten aluminum is released at the cathode, and the oxide ions released at the anode oxidize it to a *nonequilibrium mixture* composed mainly of carbon dioxide; the overall reaction is:

$$\tfrac{2}{3}Al_2O_3 + C \rightarrow CO_2 + \tfrac{4}{3}Al.$$

If this cell were to operate reversibly, the e.m.f. would depend on the free energy change for the process, which is $\Delta G^0(CO_2) - \Delta G^0(\tfrac{2}{3}Al_2O_3)$. At 1240 K, this is equal to $-396 - (-854) = +458$ kJ. Four electrons are involved in the discharge of two oxide ions, so $E^0 = 458/4(96.485) = 1.19$ V. This e.m.f. must be exceeded to produce aluminum at a measurable rate. In fact, the cells are operated at about 5 V. Notice that carbon cannot itself reduce Al_2O_3 at the operating temperature, but it does help. On the other hand, the use of a nonreacting anode would require greater electrical energy; the process as a whole would be unnecessarily expensive.

10.3.3   THERMAL REDUCTION OF MAGNESIA

Magnesium is the sixth most abundant metal in the earth's crust, but is exceedingly reactive. It finds wide use as a light alloying material, and is extensively used in the aircraft industry. Although traditionally produced by electrolytic methods, it requires between 8 and 9 kwh of electricity per pound of metal. Thermal methods are now widely used.

During the 1939–1945 war, the simple carbon reduction process was operated. The process is interesting, although obsolete:

$$2MgO(s) + 2C(s) \rightleftarrows 2CO(g) + 2Mg(g).$$

From the oxide chart (Figure 10.1) it is seen that a temperature of more than 2000 K is required to obtain negative values of ΔG^0. The cost of operating at such temperatures was high and, in addition, separation of the gaseous products was awkward. On cooling the mixture, the back reaction occurred, and so shock cooling with hydrogen was resorted to. This gave a finely powdered product, which inflamed on contact with air. Demand for magnesium was so high at that time, however, that even this inelegant process was pressed into service. Examination of the oxide chart shows that silicon, although not a good reducing agent at high temperatures, is better than carbon below about 1750 K. Having said this, it still does not show great promise in the case of magnesium. However, by making two vital modifications, Pidgeon [8] was able to show that silicon reduction was an economic possibility; it is now widely used wherever electrical energy for electrolysis is expensive. The first improvement involves the silica product of the main reaction:

$$2MgO + Si \rightarrow 2Mg + SiO_2. \tag{10.9}$$

Dolomite, $MgCO_3 \cdot CaCO_3$, occurs widely; on calcining, it yields the mixed oxide. On using this in place of MgO, the activity of the silica product is much reduced by formation of calcium silicate, $CaSiO_3$. This helps the formation of magnesium, which is gaseous at the temperature of commercial operation, 140 K. The silicate formation reaction:

$$CaO + SiO_2 \rightarrow CaSiO_3$$

contributes approximately −89 kJ to ΔG^0 for the overall process, which is

$$CaO + 2MgO + Si \rightarrow CaSiO_3 + 2\ Mg. \tag{10.10}$$

Using the oxide chart, ΔG^0 for Equation 10.9 is found to be 230 kJ, and so for Equation 10.10:

$$\Delta G^0_{1450} = +230 - 89 = +141\ kJ$$

The second modification concerns the pressure. Although ΔG^0_{1450} is +141 kJ, magnesium is the only gas, and can be pumped off if the total pressure is low enough. We can calculate ΔG for the actual process by using the general reaction isotherm:

$$\Delta G = \Delta G^0 + RT \ln Q.$$

The reaction mixture, which is calcined dolomite, ferrosilicon (75% Si) and 2% calcium fluoride as catalyst, is sealed into steel retorts that are evacuated to a pressure of about 2×10^{-5} bar. On heating to 1180°C, the Mg pressure rises to about 5×10^{-4} bar and, because $Q = p^2_{Mg}$ (all other activities being unity):

TABLE 10.2
Typical Analysis of Magnesium Obtained by the Pidgeon Process

Impurity	Si	Al	Fe	Cu	Mn	Ni
Content (ppm)	40	40	20	10	10	5

$$\Delta G = +141{,}000 + (8.314)(1453)\ln\left(5 \times 10^{-4}\right)^2 \text{ J}$$

$$= +141 - 184 = -43 \text{ kJ}.$$

Because the magnesium is distilled, the product is extremely pure. A typical analysis is shown in Table 10.2. This process depended for its development on a clear understanding of free energies and equilibria, and adequately demonstrates the power of such methods.

10.3.4 TITANIUM AND THE KROLL PROCESS

Ellingham diagrams are available for many types of reaction, such as the formation of carbides, nitrides, oxides, sulfides, chlorides, and fluorides [9], and the dissociation of sulfates and carbonates [10]. Figure 10.3 shows a simplified Ellingham diagram for chlorides, based on 1 mol of chlorine, Cl_2. It is impressive not only that chlorination reactions may be studied, but that information from two or more diagrams may be combined together.

Titanium is superbly corrosion-resistant, and has the best strength-to-weight ratio between 200 and 400°C of any commercially available metal. It is expensive to produce because methods discussed so far cannot be used. Thus, carbon reduction of the oxide leaves some titanium carbide as impurity, and electrolytic processes meet with

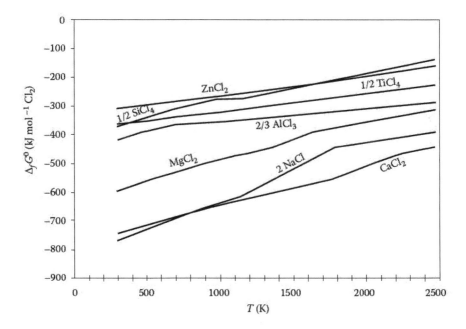

FIGURE 10.3 Ellingham diagram for the formation of a selection of chlorides, based on 1 mol of chlorine. (Based on results generated by FactSage [6] using the SGTE Pure Substance Database.)

the difficulty that $TiCl_4$ is nonconducting. In fact, titanium is formed [11] from rutile, TiO_2, by a two-stage process of chlorination to form $TiCl_4$ followed by reduction of this intermediate by a reactive metal. The reactions are:

$$TiO_2 + 2C + 2Cl_2 \rightarrow TiCl_4 + 2CO \qquad (10.11)$$

and

$$TiCl_4 + 2Mg \rightarrow 2MgCl_2 + Ti. \qquad (10.12)$$

Equation 10.11 involves the oxides of titanium and carbon, and the chloride of titanium; thus we shall make use of both the chloride and the oxide diagrams (Figures 10.1 and 10.3). The reaction takes place between 1000 and 1250 K. At 1100, ΔG^0 for Equation 10.11 is:

$$\Delta G^0_{1100} = 2\Delta G^0\left(\tfrac{1}{2}TiCl_4\right) + \Delta G^0(2CO) - \Delta G^0(TiO_2)$$

$$= 2(-314) + (-416) - (-747)$$

$$= -297 \text{ kJ}.$$

Notice that each value of ΔG^0 refers to 1 mol of O_2 or Cl_2 as appropriate. Notice also that were it not for the reducing effect of the carbon, the reaction would not be possible. (There are several metal oxides, for example, those of copper, iron and zinc, which may be chlorinated by chlorine alone.) $TiCl_4$ is a volatile, essentially covalent liquid with a boiling point of 136°C, which must be of high purity if a high quality titanium product is to be obtained.

The magnesium reduction process, represented by Equation 10.12, was developed by Kroll. Figure 10.3 shows that $TiCl_4$ is reducible by aluminum, magnesium, sodium, or calcium. The choice is governed not only by the cost of the metal, but also by the ease with which the titanium metal may be separated from the other products. Magnesium was chosen initially. To molten magnesium in a steel pot, kept under an inert gas, titanium tetrachloride is slowly added. The reaction is strongly exothermic ($\Delta H = -484$ kJ) and the temperature is kept below 1150 K. From the diagrams, $\Delta G^0 = -318$ kJ. $MgCl_2$ is molten at this temperature; it is run off from time to time. The titanium "sponge" is porous and contains $MgCl_2$ and unreacted magnesium that must be removed by leaching with dilute acid.

Alternatively, a similar procedure is followed by ICI who use liquid sodium metal. A disadvantage is that there is a very narrow range of temperature between the melting point of sodium chloride and the boiling point of sodium. On the other hand liquid sodium is more easily piped, and the leaching may be done by water. The free energy change is substantially more negative than that using magnesium, although this confers no advantage on the process.

10.3.5 SILICON METAL PRODUCTION

Silicon is the second most abundant element in the earth's crust after oxygen, and in natural form it is almost exclusively combined with oxygen as silicon dioxide and silicates. Silicon metal is produced from quartz reacted with reduction materials such as coal, coke, charcoal, and wood chips at very high temperatures with furnace electrodes.

The reaction of sand with carbon is the source of silicon. In a high temperature electric arc, we may assume the simple reaction:

$$SiO_2(s) + 2C(s) \rightleftarrows Si(l) + 2CO(g).$$

This scheme seems rather unattractive–two of nature's seemingly most stable materials in a chemical reaction to reduce sand and oxidize carbon. Intuitively, this scheme seems attractive, too. We learned that the universe seems to want to go to a disordered state of what we called high entropy. Gases, such as CO, represented ideal products of chemical reactions. Gases disperse, gases can have a huge number of configurations, gases are highly disordered, and gases have high entropy.

But we learned that if a process had high positive enthalpy and was not spontaneous at room temperature, we could only make the process go forward if the entropy was positive and we raised the temperature. It is the free energy, ΔG, that must be negative if we want a process to proceed.

From Appendix III at 298 K:

$$SiO_2(s) + 2C(s) \rightleftarrows Si(s) + 2CO(g)$$

$\Delta_r H^0_{298} = 2(-110.5) - (-910.9) = 689.9 \text{ kJ mol}^{-1}$ – highly unfavored

$\Delta_r S^0_{298} = 18.8 + 2(197.7) - 41.5 - 2(5.7) = 361.3 \text{ J mol}^{-1} \text{ K}^{-1}$

$\quad = 0.361 \text{ kJ mol}^{-1} \text{ K}^{-1}$ – highly favored but is $\Delta_r G^0_{298}$ what counts for spontaneity?

$\Delta_r G^0_{298} = 2(-137.2) - (-856.5) = 582.1 \text{ kJ mol}^{-1}$ – highly unfavorable at 298 K.

Well, what temperature would be required to, let us say, reach equilibrium in this system? At equilibrium, $\Delta G = 0$. Because all of the four substances are pure and the process takes places at atmospheric pressure, all activities are 1 and $\Delta G^0 = \Delta H^0 - T\Delta S^0 = 0$.

So let us calculate a theoretical temperature for an equilibrium, assuming that ΔH^0 and ΔS^0 are temperature independent:

$$0 = 689.9 - T(0.361)$$

$$T = 1911 \ K.$$

From the Ellingham diagram it can be seen that CO is more stable than SiO_2 at higher temperatures.

The electric arc furnace is what is required for this process, and the process will only go to equilibrium. Thus the silicon is certain to be contaminated with the two other solid materials, sand and carbon. However this temperature is far below the temperature known for this process, which is about 2200 K.

In Chapter 11, with the aid of a modern computer program, FactSage, we will see that the reaction is much more complex than the simple reaction above.

10.4   SUMMARY

This chapter has presented a set of variations on the theme of free energy changes. We have seen how accurate values of ΔG^0 may be derived at any temperature, how this data may be summarized either as f.e.f. or in terms of Ellingham diagrams, and how the data may be applied in a few instances. In virtually all cases, we have seen that activity can simplify the calculations of equilibrium constants and that allowances can always be made for nonideal behavior, assuming that activity coefficient data are available. Complete thermodynamic data have been published for relatively few compounds, however, and there are, for example, many common organic compounds for which only an enthalpy of formation has been determined. As more complete information is circulated, the number of applications of chemical thermodynamics will continue to increase. With the aid of thermochemical databases, calculations can often be carried out in a matter of minutes at the terminal.

REFERENCES

1. Lewis, G. N., and M. Randall. 1961. *Thermodynamics*. 2nd ed., Chap. 15 and Appendix 7. New York, NY: McGraw-Hill.
2. Shell Development Company. Emeryville, CA.
3. Bakshi, Y. M., A. I. Gel'bshtein, and M. I. Temkin. 1960. Additional data on the equilibrium in the synthesis of ethyl alcohol. *Dokl Akad Nauk SSSR* 132:157–159.
4. Ellingham, H. J. T. 1944. Reducibility of oxides and sulfides in metallurgical processes. *J Soc Chem Ind* 63:125–133.
5. Ives, D. J. G. 1960. *Principles of the Extraction of Metals*. Monograph 3. Royal Institute of Chemistry, London.
6. FactSage©, one of the largest fully integrated database computing systems in chemical thermodynamics in the world, was introduced in 2001 and is the fusion of the FACT-Win/F*A*C*T and ChemSage/SOLGASMIX thermochemical packages. FactSage is the result of over 20 years of collaborative efforts between Thermfact/CRCT (Montreal, Canada) www.crct.polymtl.ca and GTT-Technologies (Aachen, Germany) www.gtt-technologies.de
7. McGuire, M. F. 2008. *Stainless Steel for Design Engineers*. Cleveland, Ohio: ASM International.
8. Pidgeon, L. M. 1946. *Trans Can Min Inst* 49:621–635; Pidgeon, L. M., and J. A. King. 1948. Vapor pressure of magnesium in the thermal reduction of magnesium oxide by ferrosillicon. *Disc Faraday Soc* 4:197–206.
9. Smithells, C. J. 1955. *Metals Reference Book*. Vol. 2. London: Butterworths.

10. Hopkins, D. W. 1954. *Physical Chemistry and Metal Extraction*, 192, London: Garnett Miller.
11. Gray, J. J., and Carter A. 1958. *Chemistry and Metallurgy of Titanium Production*. Royal Institute of Chemistry, Lectures, Monographs, and Reports, 1.

PROBLEMS

10.1   Assuming that ΔH^0 and ΔS^0 are temperature-invariant, make an approximate calculation of K for the reaction: $CO + 2H_2 \rightarrow CH_3OH$ at 350°C, using tabulated data.

10.2   An accurate expression for ΔG^0 (in J mol^{-1}) for the methanol synthesis (see previous question) is:

$$\Delta G^0 = -73{,}350 + 76.11 T \ln T - 0.0590 T^2 - 252.7 T.$$

(a) Find K at 350°C.

(b) First, assuming that $K = K_p$ ($K_\gamma = 1$), calculate the partial pressure of methanol in an equilibrium mixture held at 250 bar. The reactants are in the stoichiometric proportion of 1:2 CO/H$_2$. What proportion of CO is converted?

(c) Using generalized fugacity charts, K_γ is found to be 0.41 under the actual operating conditions. Calculate K_p in this case, and thence the equilibrium pressures of CO, H$_2$ and CH$_3$OH. What is the CO conversion?

10.3   In some nuclear reactors, UO$_2$ fuel elements in stainless steel cans are subjected to high rates of "burning"–that is, substantial quantities of fission products are formed. As a result, appreciable amounts of O$_2$ are liberated (from UO$_2$), which will then react either with the fission products (zirconium, molybdenum, cerium, etc.) or with the stainless steel of the can. Decide, with the help of Figure 10.2, whether zirconium and molybdenum are oxidized in preference to the steel ($T = 750°C$). (In fact sufficient reactive elements are formed to take up the oxygen.)

10.4   Using Figure 10.2, decide at what temperature, if any, UO$_2$ fuel oxidizes the stainless steel can (see previous example).

10.5   Magnesium has been produced by the following reaction:

$$\tfrac{4}{3}Al(l) + 2\,MgO \rightarrow 2\,Mg(g) + \tfrac{2}{3}Al_2O_3.$$

Process conditions commonly used were $T = 1200°C$, and $P = 0.02$ bar. First, calculate ΔG^0 for the process, from the oxide Ellingham diagram, then, use the reaction isotherm to calculate ΔG under operating conditions.

10.6   Calcium is produced by the electrolysis of molten calcium chloride. What is the minimum (equilibrium) voltage needed for electrolysis at 1048 K, the melting point of the chloride. (Use Figure 10.3.)

10.7   Manganese is produced by carbon reduction of pyrolusite, MnO_2, which occurs in two stages: $MnO_2 \rightarrow MnO \rightarrow Mn$. Determine the lowest temperature for the second stage of the process.

10.8   Pure silicon may be prepared by the reaction:

$$SiCl_4 + 2Zn \rightarrow 2ZnCl_2 + Si.$$

Find from the chloride diagram ΔG^0 for this reaction at 1000°C. What is K approximately at this temperature?

10.9   Magnesium is often produced by electrolysis of a molten mixed chloride electrolyte containing $MgCl_2$, at 750°C. Molten magnesium is released around the cast steel cathodes. Use an Ellingham diagram to determine the equilibrium potential necessary for decomposition.

In fact, 7.54 kwh are used to produce 1 lb (454 g) of metal. What is the actual potential used, and why is such an excess voltage necessary?

11 Computational Thermochemistry

Computer-assisted thermochemistry is a tool that can be applied in many fields today. In particular, with the aid of reliable thermochemical databases and appropriate application software, optimum operating temperatures, reacting amounts, and/or gas pressures necessary to obtain a product of the required purity can be calculated. Costly and time-consuming experimental work can thereby be reduced considerably [1].

The calculation of thermochemical equilibrium states is the basis for the solution of many problems in process engineering or materials science. The calculations are carried out by minimization of the Gibbs energy of the system under consideration, taking proper account of the global parameters, system composition, temperature, and total pressure. One of the most up-to-date computer programs for this purpose is FactSage (Chapter 10, Ref. 6).

FactSage is the single successor to both FACT-Win and ChemSage. It combines the powerful calculational capabilities of ChemSage developed by GTT and the very versatile and user friendly graphical user interface generated by FACT/CRCT into one of the most powerful and versatile interactive programs available in the field of computational thermochemistry.

In conjunction with critically evaluated thermochemical data (e.g., SGTE database) [2], this program permits a reliable description and, in consequence, optimization of process variables or material properties. Several examples will be given below. It is worth noting that although these examples stem from very different technical fields, the general approach to problem solving is the same in all cases because of the same underlying physicochemical principles.

11.1 CALCULATION OF AN ADIABATIC FLAME TEMPERATURE

The flame temperature in Example 3.3 will be calculated with the program FactSage. In the first calculation, the assumption is that the following reaction takes place:

$$C_2H_2 + 3N_2O \rightarrow 2CO + H_2O + 3N_2.$$

Table 11.1 shows the results of this calculation. The result of 4404 K is very close to that given in Example 3.3 (4426 K). The small difference is due to other C_P functions in the database that was used in ChemSage (the older version of FactSage) for the previous edition. A full interpretation of the output of Table 11.1 is given when the second calculation of the flame temperature is done.

TABLE 11.1

Calculation of the Adiabatic Flame Temperature of the Reaction $C_2H_2 + 3N_2O \rightarrow 2CO + H_2O + 3N_2$

$T = 4404.44$ K
$p = 1.00000$ bar
$V = 2.1972 \times 10^3$ L

Reactants	Amount (mol)	Temperature (K)	Pressure (bar)
C_2H_2/gas	1.0000	298.15	1.0000
N_2O/gas	3.0000	298.15	1.0000

	Equilibrium Amount (mol)	Mole Fraction	Fugacity (bar)
Phase: Gas			
N_2	3.0000	5.0000×10^{-1}	5.0000×10^{-1}
CO	2.0000	3.3333×10^{-1}	3.3333×10^{-1}
H_2O	1.0000	1.6667×10^{-1}	1.6667×10^{-1}
Total	6.0000	1.0000	1.0000

ΔH (J)	ΔS (J/K)	ΔG (J)	ΔU (J)	ΔA (J)	ΔV (L)
0.0000	8.9910×10^2	-7.5699×10^6	-2.0981×10^5	-7.7797×10^6	2.0981×10^3

It is also possible with the FactSage program to calculate the adiabatic flame temperature without neglecting possible by-products. In this case, the reaction will be:

$$C_2H_2 + 3\,N_2O \rightarrow \cdots$$

Table 11.2 shows the results of this calculation. The table consists of a header A; the reactant section B; equilibrium phase sections, C and E; and a case-dependent output section, F.

A: The header contains values for the temperature, pressure, and volume of the system. In this particular case, temperature is a calculated result based on the constraint of adiabatic conditions $(H_2(T_2, \ldots) - H_1(T_1, \ldots) = 0)$.

B: The reactants are given by name and phase (where appropriate), and their overall amounts (column 2). If, as here, there is an extensive property calculation, the initial temperature (column 3) and pressure (column 4) are also defined.

C and E: These sections contain information on the equilibrium phases of the system. Column 1 gives the names; column 2 gives the equilibrium amounts of the species; column 3 shows the phase internal concentrations for the solution phases; and column 4 contains the fugacities in the case of gases (C) and the equilibrium activity in the case of condensed phases (E). In this case, only pure water and carbon are possible condensed phases.

F: In this output table, the values in the case-dependent output section relate to the extensive property changes of the reaction between the input substances in section B and the equilibrium substances in sections C and E. Note that the input value of zero in column 1 for the enthalpy change indicates that the adiabatic condition was

TABLE 11.2

Calculation of the Adiabatic Flame Temperature of the Reaction:

$C_2H_2 + 3N_2O \rightarrow \cdots$

A	$T = 3255.09$ K $p = 1.00000$ bar $V = 1.7760 \times 10^3$ L			

	Reactants	Amount (mol)	Temperature (K)	Pressure (bar)
B	C_2H_2/Gas	1.0000	298.15	1.0000
	N_2O/Gas	3.0000	298.15	1.0000

		1	2	3	4
	Phase: Gas				
	N_2	2.9685	4.5238×10^{-1}	4.5238×10^{-1}
	CO	1.7959	2.7368×10^{-1}	2.7368×10^{-1}
	H	4.9370×10^{-1}	7.5236×10^{-2}	7.5236×10^{-2}
	H_2	3.5402×10^{-1}	5.3951×10^{-2}	5.3951×10^{-2}
	H_2O	3.1073×10^{-1}	4.7353×10^{-2}	4.7353×10^{-2}
C	CO_2	2.0408×10^{-1}	3.1100×10^{-2}	3.1100×10^{-2}
	HO	1.7672×10^{-1}	2.6931×10^{-2}	2.6931×10^{-2}
	O	1.4493×10^{-1}	2.2087×10^{-2}	2.2087×10^{-2}
	NO	6.2634×10^{-2}	9.5449×10^{-3}	9.5449×10^{-3}
	O_2	5.0406×10^{-2}	7.6814×10^{-3}	7.6814×10^{-3}
	N	2.7802×10^{-4}	4.2368×10^{-5}	4.2368×10^{-5}
	HO_2	3.6570×10^{-5}	5.5729×10^{-6}	5.5729×10^{-6}
	HN	1.0355×10^{-5}	1.5780×10^{-6}	1.5780×10^{-6}
	Total	6.5620	1.0000	1.0000

			(mol)		Activity
E	H_2O	T	0.0000		8.6555×10^{-5}
	C		0.0000		1.8189×10^{-6}

	ΔH (J)	ΔS (J/K)	ΔG (J)	ΔU (J)	ΔA (J)	ΔV (L)
F	0.0000	9.2782×10^2	-5.6196×10^6	-1.6768×10^5	-5.7873×10^6	1.6768×10^3

Note: Data on one species identified with "*T*" have been extrapolated.

used as a constraint, that is, *extensive property target*, to find the temperature value shown in A.

Species with an equilibrium amount smaller than 10^{-5} have been omitted. The magnitude of activity values below unity is a measure of thermochemical instability. A value close to unity indicates that even small changes in temperature, pressure, etc., can cause a phase to become stable. It is remarkable how close the calculated flame temperature of 3255 K is to the practical value of 3230 K. The main reason why the result of Example 3.3 differs so much from the practical value is neglect of the "by-products" rather than neglect of radiation losses.

TABLE 11.3
Calculation of the Equilibrium in the Fe-B-C-Mn-Al-O-Nb System at 1223 K, Including the Niobium Carbonitride Phase

$T = 1223.0$ K
$p = 1.0000$ bar
$V = 0.0000$ L

Reactants	Weight (g)
Mn/solid-Fe/	1.2000
S/solid-Fe/	2.0000×10^{-2}
B/solid-Fe/	5.0000×10^{-3}
Al/solid-Fe/	4.0000×10^{-2}
Nb/solid-Fe/	3.5000×10^{-2}
O/solid-Fe/	1.0000×10^{-3}
C/solid-Fe/	2.5000×10^{-1}
N/solid-Fe/	2.0000×10^{-2}
Fe/solid-Fe/	9.8429×10^{1}

	Equilibrium Amount (mol)	Pressure (bar)	Fugacity (bar)
Phase: Gas			
N_2	0.0000	6.0288×10^{-4}	6.0288×10^{-4}
CO	0.0000	3.6891×10^{-6}	3.6891×10^{-6}
Mn	0.0000	1.3277×10^{-7}	1.3277×10^{-7}
Cs	0.0000	1.7697×10^{-10}	1.7697×10^{-10}
Fe	0.0000	1.0575×10^{-10}	1.0575×10^{-10}
Total	0.0000	6.0670×10^{-4}	

	Gram	Weight Fraction	Activity
Phase: Nb Carbonitride			
NbC	3.5953×10^{-2}	9.1284×10^{-1}	9.1432×10^{-1}
NbN	3.4330×10^{-3}	8.7164×10^{-2}	8.5676×10^{-2}
Total	3.9386×10^{-2}	1.0000	1.0000

	Gram	Weight Fraction	Activity
Phase: Solid Fe			
Fe	9.8429×10^{1}	9.8571×10^{-1}	9.7657×10^{-1}
Al	1.4774×10^{-2}	1.4795×10^{-4}	3.0339×10^{-4}
B	6.4861×10^{-5}	6.4954×10^{-7}	3.3243×10^{-6}
C	2.4588×10^{-1}	2.4624×10^{-3}	1.1343×10^{-2}
Mn	1.1657	1.1674×10^{-2}	1.1757×10^{-2}
N	6.4446×10^{-4}	6.4539×10^{-6}	2.5494×10^{-5}
Nb	1.8014×10^{-4}	1.8040×10^{-6}	1.0744×10^{-6}
O	3.6926×10^{-11}	3.6979×10^{-13}	1.2788×10^{-12}
S	4.8712×10^{-6}	4.8782×10^{-8}	8.4178×10^{-8}
Total	9.9856×10^{1}	1.0000	1.0000

(*continued*)

TABLE 11.3 (Continued)
Calculation of the Equilibrium in the Fe-B-C-Mn-Al-O-Nb System at 1223 K, Including the Niobium Carbonitride Phase

	Gram	Activity
MnS	5.4254×10^{-2}	1.0000
AlN	3.6613×10^{-2}	1.0000
BN	1.1329×10^{-2}	1.0000
Al_2O_3	2.1243×10^{-3}	1.0000
C	0.0000	1.9258×10^{-1}
FeB	0.0000	1.9259×10^{-2}
Fe_2B	0.0000	1.6090×10^{-2}
Mn	0.0000	1.4936×10^{-2}
Nb_2C	0.0000	3.9978×10^{-4}
MnB	0.0000	2.8785×10^{-4}
$NbFe_2$	0.0000	2.2720×10^{-4}
NbB_2	0.0000	9.1932×10^{-5}
B	0.0000	2.9421×10^{-5}
Nb	0.0000	6.7450×10^{-6}
$FeO \cdot Al_2O_3$	0.0000	2.9835×10^{-6}
Nb_2N	0.0000	2.3191×10^{-6}
Al	0.0000	4.9043×10^{-7}
Mn_4N	0.0000	2.5902×10^{-7}
MnB_2	0.0000	3.4273×10^{-8}
S	0.0000	2.9168×10^{-8}

11.2 PRECIPITATION OF CARBIDE AND NITRIDE PHASES FROM DILUTE SOLUTION IN ALLOY STEEL

The presence of carbide and nitride precipitates in alloy steels can have a beneficial effect on the mechanical properties of the steels concerned [3]. However, the amounts, morphology, and distribution of the precipitated phases must be carefully controlled to achieve the properties required. Because the presence of hard precipitates in a steel during hot-rolling operations can result in damage both to the rollers and to the steel, it is important that information be available on the ranges of temperature and composition in which precipitated phases are stable. For this reason, and also to achieve the desired precipitation characteristics using the minimum amounts of expensive precipitating elements such as niobium, titanium, vanadium, etc., it is helpful to carry out prior calculations of the stability of precipitates in steels of different compositions.

Table 11.3 presents the results of an equilibrium calculation relating to the precipitation of a number of phases, including the niobium carbonitride phases, from

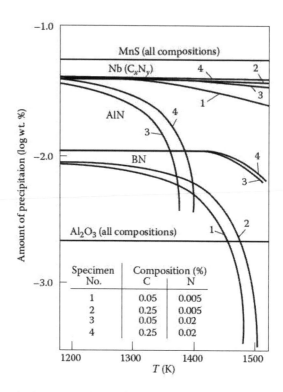

FIGURE 11.1 Calculated precipitated behavior in steel containing 1.2% Mn, 0.02% S, 0.005% B, 0.04% Al, 0.035% Nb, 0.001% O, 0.05–0.25% C, and 0.005–0.02% N.

dilute solution in an austenitic steel at 1223 K. The calculation shows that at this temperature, the Nb (N, C) phase itself is stable, together with AlN, BN, and MnS, as other precipitated phases. By carrying out a series of such calculations for a number of temperatures for the given steel, changes in the amounts of the precipitated phases can be determined, as well as the composition of the carbonitride phase. This is illustrated for the present steel in Figure 11.1.

11.3 CVD PRODUCTION OF ULTRAPURE SILICON

A simple, but representative, example of the application of FactSage to the investigation of chemical vapor deposition (CVD) processes is the production of ultrapure silicon by the thermal decomposition of $SiHCl_3$ gas [3]. In this process, it is of technological and ecological importance to establish the optimum temperatures for the maximum yield of pure silicon from the decomposition reaction.

Figure 11.2 illustrates the calculated silicon yield (mol) from 1 mol of $SiHCl_3$ gas as a function of temperature at 1 atm pressure. The plot shows that an optimum yield can be expected at temperatures around 1100 K. This is in close agreement with experimental observations. The corresponding composition of the gas phase as a function of temperature is illustrated in Figure 11.3.

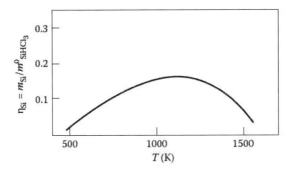

FIGURE 11.2   Si yield (mol) produced by the dissociation of 1 mol $SiHCl_3$ gas, as a function of temperature.

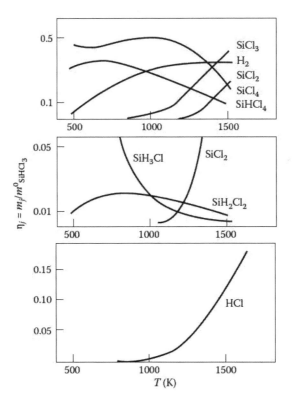

FIGURE 11.3   Gaseous species resulting from the dissociation of 1 mol $SiHCl_3$ gas, as a function of temperature.

11.4   PROCESSING OF WASTES FROM THE ALUMINUM ELECTROLYTIC FURNACE

After a sufficient length of time in operation, the electrolytic cell used to produce aluminum must be renewed and the waste materials it contains treated to produce solid phases that can either be safely dumped or recycled [3]. A fluidized bed reactor is suitable for the combustion of the waste products of the cell, using humidified air.

The significant amounts of fluorine in the waste materials in question, which would cause serious environmental problems if the waste were dumped in untreated form, can be recovered as a result of the formation of HF gas during the combustion process. The recovery can be optimized by establishing the influence on HF yield of parameters such as temperature, humidity of the combustion air, and enrichment with oxygen.

Using typical waste material analyses, calculation has been made of the variation of HF partial pressure with H_2O/O_2 ratio for various N_2/O_2 concentrations and selected combustion temperatures. The results are illustrated in Figure 11.4a and b. The calculated enthalpy change associated with the combustion reaction in each case is also shown in Figure 11.5 for a temperature of 1400 K. With the aid of such

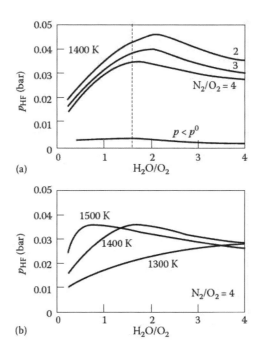

FIGURE 11.4   (a) Partial pressure of HF gas produced by combustion of waste materials from the aluminum electrolytic furnace as a function of H_2O/O_2 ratio, for different N_2/O_2 concentrations at 1400 K. (b) Partial pressure of HF gas produced from the aluminum electrolytic furnace as a function H_2O/O_2 ratio, for selected combustion temperatures.

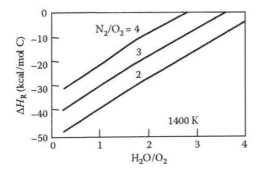

FIGURE 11.5 Enthalpy of the combustion reaction as a function of H_2O/O_2 ratio, for different N_2/O_2 concentrations at 1400 K.

information, the most economical and safe conditions for carrying out the waste combustion process can be selected.

11.5 PRODUCTION OF METALLURGICAL-GRADE SILICON IN AN ARC (OVEN) FURNACE

In Section 10.3.5 we studied the reaction:

$$SiO_2(s) + 2C(s) \rightleftarrows Si(l) + 2CO(g).$$

However, the calculated temperature (1911 K) at which $\Delta G = 0$ is far below the temperature known for this process, which is about 2200 K. If quartz is permitted to react freely with carbon at a given total pressure of 1 bar FactSage can generate a table similar to Table 11.3 for temperatures between 1600 and 3000°C.

Without going into detail by giving all data, a great number of gaseous species and also stoichiometric condensed phases can be formed. In the gas phase, SiO is an essential species. In the reaction above, however, no SiO was included, so this reaction cannot be correct. In the present case, SiO is permitted to form together with SiC, so the chemistry is correct. However, the temperature at which silicon is formed is near 2900 K and the yield is not more than 50%. In the real process, the reaction temperature is about 2200 K. So the calculation of single full equilibrium states for a total pressure of 1 bar and for a series of different temperatures does not agree with experimental data.

What is overlooked is the fact that in an arc furnace, there is a temperature gradient and that cold condensed matter is fed through the top of the furnace, falling downward, while hot gases flow rapidly upward [4]. On their way, the substances meet and exchange heat or even react with each other.

The furnace consists of a carbon-lined crucible into which are suspended three prebaked carbon electrodes. Very high electrical current is passed down the electrodes and a regulated arc is formed between the tip of the electrode and the carbon base of the crucible (hearth). (See the photo on the front cover of this book.) A steep temperature gradient extends from the zone of the arc (>3300 K) below the electrode to the top surface of the charge mix (approximately 1000 K).

The furnace charge is a mixture of quartz, carbonaceous reducing agents (charcoal, coal, petroleum coke), and wood chips. If the silicon arc furnace is divided into four separate parts that are heat balance–controlled and, therefore, at different temperatures, the results of the calculations with a dedicated process model generated using the Rapid Process Application tool SimuSage [5] can be given in the flow diagrams in Figures 11.6 and 11.7. In stages I to III, there are adiabatic conditions.

The reactor can be split into two distinct zones that are governed by two separate processes. At the bottom of the furnace is the main reaction zone (metal production zone) where, at temperatures exceeding 2100 K, silicon is produced according to the equation:

$$SiO_2 + SiC \rightarrow Si(l) + SiO(g) + CO(g).$$

This silicon is drained from the furnace via a taphole into ladles in which it is refined and transferred to the casting area. In the top zone, the upflowing $CO(g)$ reacts with the downflowing carbon according to the equation:

$$C + SiO(g) \rightarrow 1/3\ SiO_2 + 2/3\ SiC + 1/3\ CO(g).$$

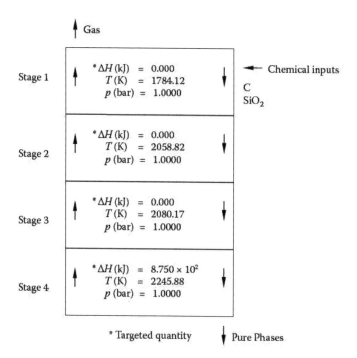

FIGURE 11.6   Flow diagram for the silicon arc furnace.

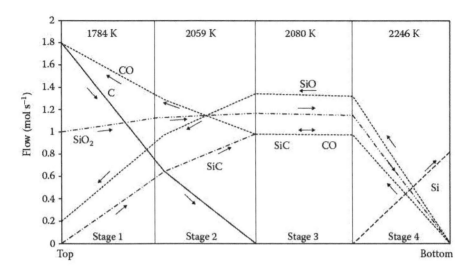

FIGURE 11.7   The materials flow for the calculated steady state.

Silicon monoxide that fails to react within the furnace oxidizes in the atmosphere to form SiO_2 (a dustlike material called amorphous silica fume). The silica fume is vented away for collection in a large filtration facility (baghouse) as a by-product of the silicon production.

$$2SiO + O_2 \rightarrow 2SiO_2.$$

These data agree fairly with the experimental data from the silicon arc furnace of KemaNord at Ljungavrek, Sweden.

11.6   SUMMARY

The thermochemical SGTE database enables critically evaluated data and interactive calculation programs, in particular FactSage, to be combined in rapid theoretical thermochemical investigations of materials problems of all types. The results can be obtained directly from the user's PC in the graphical or tabular form he or she requires. The calculations also provide basic information for minimization of energy and material wastage.

The present contributions show how carefully chosen thermochemical data, in conjunction with the computer program FactSage, can assist in the understanding of processes in silicon metallurgy. Furthermore, we have shown that the same software can also be used in combustion calculations or for the understanding of phase relations in hard-metal alloys. In all cases rapid and reliable answers for multicomponent, multiphase equilibria are obtained.

REFERENCES

1. Köningsberger, E., E. Schuster, H. Gamsjäger, C. God, K. Hack, M. Kowalski, and P. J. Spencer. 1992. Thermochemical data and software for the optimization of processes and materials. *Netsu Sokutei* 19(3):135–144.

2. SGTE, Scientific Group Thermodata Europe, Domaine Universitaire de Grenoble, 1001 avenue Centrale BP 66, 38402 Saint Martin d'Hères Cedex, France. http://www.sgte .org.

3. Spencer, P. J., and K. Hack. 1990. The solution of materials problems using the thermochemical databank system THERDAS. *Swiss Chem* 12:69–73.

4. Hack, K. 2008. *The SGTE Casebook—Thermodynamics at Work*. Boca Raton, FL: CRC Press.

5. Petersen, S., K. Hack, P. Monheim, and U. Pickartz. 2007. SimuSage - the component library for rapid process modelling and its application. *Int J Mater Res* 98(10): 946–953.

Appendix I
The Twenty Most Useful Equations

If thermochemical data are the flesh and blood of chemical thermodynamics, then it is true to say that the bone structure is made up of a small number of defining and operating equations. To make such a selection is necessarily arbitrary; to present it is the best way of demonstrating the underlying structure of the subject.

$$\Delta U = q_{to} + w_{on}; \text{ defining equation for } U \tag{2.1}$$

$$H = U + PV; \text{ defining equation for } H \tag{2.5}$$

$$\Delta H = \Delta U + P\Delta V \quad (P \text{ constant}) \tag{2.7}$$

$$\Delta H = \Delta U + \Delta nRT \quad (P,T \text{ constant}) \tag{2.9}$$

where n is the number of moles of gas.

$$(d(\Delta H)/dT)_P = \Delta C_P; \text{ Kirchhoff's equation} \tag{3.4}$$

$$dS = dq/T \quad (\text{reversible change}) \tag{5.1}$$

$$dS > dq/T \quad (\text{irreversible change});$$

formal definitions of entropy

$$\Delta S = nR \ln V_2/V_1 \quad (T \text{ constant, ideal gas}) \tag{5.3}$$

$$\Delta S = \int_{T_1}^{T_2} \frac{C_p}{T} dT \quad (P \text{ constant}) \tag{5.5}$$

$$S_T = \int_0^T \frac{C_p}{T} dT \tag{5.8}$$

$$G = H - TS; \text{ defining equation for } G \tag{6.1}$$

$$\Delta G = \Delta H - T\Delta S \quad (T \text{ constant}) \tag{6.3}$$

$$\Delta G = \Delta G^0 + RT \ln Q;$$

general reaction isotherm

$$(7.14)$$

$\Delta G^0 = -RT \ln K$; equilibrium isotherm.

Standard free energy change is directly linked to $\qquad$ (7.15)

equilibrium constant in activities.

$$\ln K = -\frac{\Delta H^0}{RT} + \frac{\Delta S^0}{R}; \text{ reaction isochore} \qquad (8.2)$$

$$\frac{dP}{dT} = \frac{\Delta H}{T \Delta V}; \text{ Clapeyron equation} \qquad (8.7)$$

$$\Delta G^0 = -nFE^0 \qquad (9.2)$$

$$E = E^0 + \frac{RT}{nF} \ln \frac{a_{\text{Oxidized States}}}{a_{\text{Reduced States}}} \qquad (9.6)$$

$$\Delta S^0 = nF \left(\frac{dE^0}{dT} \right)_P \qquad (9.11)$$

$$\Delta G = \Delta H + T \left(\frac{d(\Delta G)}{dT} \right)_P;$$

$$(10.2)$$

the Gibbs-Helmholtz equation

Appendix II

Fundamental Constants and Conversion Factors

Volume of 1 mol ideal gas at s.t.p.	$V = 24.790$ L mol^{-1}
Universal gas constant	$= 8.3145$ J mol^{-1} K^{-1}
	$= 0.08205$ L atm mol^{-1} K^{-1}
	$= 1.9872$ cal mol^{-1} K^{-1}
Faraday constant	$F = 96,485.3$ C mol^{-1}
	$= 23,061$ cal V^{-1} mol^{-1}
Ice point	$0°C = 273.15$ K
Avogadro constant	$N_A = 6.02214 \times 10^{23}$ particles mol^{-1}

1 J = 1 V coulomb = 0.009869 L atm

1 cal = 4.184 J (exactly) (defined) = 0.04129 L atm

1 mol = N particles

1 atm = 101,325 Pa (definition) = 1.01325 bar

1 bar = 10^5 Pa

2.3026 RT/F = 0.05916 V at 298.15 K

1 quad = 293 GW h

Appendix III

A Selection of Basic Thermodynamic Data (Data taken from SGTE databases[a])

Substance	State[b]	$\Delta_f H^0_{298}$ (kJ mol^{-1})	$\Delta_f G^0_{298}$ (kJ mol^{-1})	S^0_{298} (J mol^{-1} K^{-1})
		Aluminum		
Al	g	330.0	289.4	164.6
Al	s	0	0	28.3
Al$_2$O$_3$	s	-1676	-1582	50.9
		Barium		
Ba	g	185.0	152.9	170.2
Ba	s	0	0	62.5
Ba^{2+}	aq	-537.6	-560.8	9.62
BaCl$_2$	s	-855.2	-806.9	123.7
BaCl$_2 \cdot 2$H$_2$O	s	-1460.1	-1296.4	202.9
Ba(OH)$_2 \cdot 8$H$_2$O	s	-3342.2	-2794.4	427.0
BaSO$_4$	s	-1459	-1348	132.1
		Boron		
B	g	565.0	521.0	153.4
B	s	0	0	5.90
B	amorphous	4.40	4.21	6.53
B$_2$O$_3$	s	-1274	-1194	54.0
HBO$_2$	s	-804.6	-736.8	49.0
BF$_3$	g	-1136	-1119	254.4
B$_2$H$_6$	g	36.60	87.69	232.5
		Bromine		
Br	g	111.9	82.38	175.0
Br$^-$	aq	-121.5	-103.9	82.4
Br$_2$	l	0	0	152.2
Br$_2$	g	30.91	3.11	245.5
HBr	g	-36.29	-53.36	198.7
HBr	aq	-121.5	-103.9	82.4
		Calcium		
Ca	g	177.8	144.0	154.9
Ca	s	0	0	41.6
Ca^{2+}	aq	-542.8	-553.5	-53.1
CaCO$_3$	s	-1207	-1128	91.7
CaCl$_2$	s	-795.4	-748.8	108.4
CaO	s	-634.9	-603.3	38.1
Ca(OH)$_2$	s	-985.9	-898.2	83.4

(continued)

Substance	State[b]	$\Delta_f H^0_{298}$ (kJ mol^{-1})	$\Delta_f G^0_{298}$ (kJ mol^{-1})	S^0_{298} (J mol^{-1} K^{-1})
		Carbon		
C	g	716.7	671.2	158.1
C	graphite	0	0	5.69
CO_2	g	−393.5	−394.4	213.8
CO	g	−110.5	−137.2	197.7
CS_2	l	89.41	65.11	151.3
CS_2	g	116.7	66.59	237.9
		Chlorine		
Cl	g	121.3	105.3	165.2
Cl$^-$	aq	−167.2	−131.3	56.5
Cl_2	g	0	0	223.1
HCl	g	−92.31	−95.30	186.9
HCl	aq	−167.2	−131.3	56.5
		Copper		
Cu	g	337.6	297.9	166.4
Cu	s	0	0	33.15
Cu^{2+}	aq	64.77	65.40	−99.6
CuO	s	−155.6	−127.9	42.6
Cu_2O	s	−173.1	−150.3	92.6
$CuSO_4$	s	−770.0	−660.8	109.3
$CuSO_4 \cdot H_2O$	s	−1083	−914.8	145.1
		Fluorine		
F	g	79.38	62.28	158.7
F_2	g	0	0	202.8
HF	g	−273.3	−275.4	173.8
HF	aq	−332.6	−278.8	−13.8
		Hydrogen		
H	g	218.0	203.3	114.7
H$^+$	aq	0	0	0
H_2	g	0	0	130.7
H_2O	g	−241.8	−228.6	188.8
H_2O	l	−285.8	−237.1	70.0
H_2O_2	l	−187.9	−120.4	109.6
		Iodine		
I	g	106.8	70.18	180.8
I$^-$	aq	−55.19	−51.59	111.3
I_2	s	0	0	116.1
I_2	g	62.42	19.32	260.7
HI	g	26.50	1.70	206.6
HI	aq	−55.19	−51.59	111.3
		Iron		
Fe	s	0	0	27.3
Fe^{2+}	aq	−92.3	−91.5	−105.9

(*continued*)

Substance	State[b]	$\Delta_f H^0_{298}$ (kJ mol^{-1})	$\Delta_f G^0_{298}$ (kJ mol^{-1})	S^0_{298} (J mol^{-1} K^{-1})
FeO	s	−272.0	−251.4	60.8
Fe_2O_3	gamma	−804.0	−723.4	91.8
Fe_2O_3	hematite	−823.4	−741.5	87.5
Lithium				
Li	s	0	0	29.1
LiH	s	−90.65	−68.63	20.6
LiOH	s	−484.9	−438.9	42.8
Magnesium				
Mg	s	0	0	32.7
$MgCl_2$	s	−644.3	−594.8	89.6
$MgBr_2$	s	−526.0	−505.8	117.0
MgO	s	−601.6	−569.3	27.0
Nickel				
Ni	s	0	0	29.9
NiO	s	−240.6	−212.4	38.1
Nitrogen				
N	g	472.7	455.5	153.3
N_2	g	0	0	191.6
NH_3	g	−45.94	−16.41	192.8
NH_4Cl	s	−314.6	−373.7	95.0
NO	g	91.28	87.59	210.7
N_2O	g	81.6	103.7	220.0
NO_2	g	34.19	52.32	240.2
N_2O_4	g	11.11	99.80	304.4
NOCl	g	52.7	67.11	261.6
HCN	g	132.0	121.6	201.8
Oxygen				
O	g	249.2	231.7	161.1
O_2	g	0	0	205.10
O_3	g	141.8	162.3	239.0
Phosphorus				
P	s, white	0	0	41.1
PCl_3	g	−289.5	−270.4	311.7
PCl_5	g	−376.0	−307.0	367.2
Potassium				
K	s	0	0	64.7
K^+	aq	−252.4	−283.2	102.5
Silicon				
Si	s	0	0	18.8
SiH_4	g	34.31	56.82	204.7
SiO_2	quartz	−910.9	−856.5	41.5
Silver				
Ag	s	0	0	42.6

(*continued*)

Substance	State[b]	$\Delta_f H^0_{298}$ (kJ mol^{-1})	$\Delta_f G^0_{298}$ (kJ mol^{-1})	S^0_{298} (J mol^{-1} K^{-1})
Ag$^+$	aq	105.8	77.1	96.1
AgCl	s	−127.1	−109.8	96.2
AgClO$_2$	s	0	66.94	134.6
AgClO$_3$	s	−30.29	65.11	141.8
Sodium				
Na	s	0	0	51.3
Na$^+$	aq	−240.1	−261.9	59.0
NaCl	s	−411.1	−384.1	72.1
Na$_2$O	s	−418.0	−379.2	75.0
NaI	s	−289.6	−286.4	98.6
Sulfur				
S	g	277.2	236.7	167.8
S	s, rhombic	0	0	32.1
S$_2$	g	128.6	79.70	228.2
SOCl$_2$	l	−245.6	−203.4	215.7
SOCl$_2$	g	−212.0	−197.0	307.4
H$_2$S	g	−20.60	−33.44	205.8
SO$_2$	g	−269.8	−300.1	248.2
SO$_4^{2-}$	aq	−909.3	−744.6	20.1
Titanium				
Ti	s	0	0	30.7
TiCl$_4$	l	−801.7	−733.8	249.4
TiO$_2$	s, anatase	−933.0	−877.6	49.9
Tungsten				
W	g	851.2	809.1	174.0
W	s	0	0	32.7
WBr$_5$	g	−199.2	−213.5	461.4
Zinc				
Zn	s	0	0	41.6
Zn^{2+}	aq	−153.9	−147.0	−112.1
ZnO	s	−350.5	−320.5	43.7
ZnS	s	−205.2	−200.4	57.7
Organic Compounds				
CH$_4$	g	−74.60	−50.55	186.4
C$_2$H$_2$	g	227.4	209.8	200.9
C$_2$H$_4$	g	52.40	68.33	219.3
C$_2$H$_6$	g	−84.00	−32.05	229.2
C$_3$H$_8$	g	−103.8	−23.42	270.0
n-C$_4$H$_{10}$	g	−126.1	−17.05	310.2
iso-C$_4$H$_{10}$	g	−134.5	−20.80	294.7
C$_6$H$_6$	g	82.93	129.7	269.3
CCl$_4$	l	−128.0	−57.81	216.4

(*continued*)

Substance	State[b]	$\Delta_f H^0_{298}$ (kJ mol^{-1})	$\Delta_f G^0_{298}$ (kJ mol^{-1})	S^0_{298} (J mol^{-1} K^{-1})
CHCl$_3$	l	−134.5	−73.72	201.7
CH$_3$Cl	g	−81.87	−58.36	234.4
C$_2$H$_4$Br$_2$	g	−41.00	−12.00	327.6
CH$_2$BrCH$_2$Br	l	−85.3	−25.9	226.0
CHClF$_2$	g	−475.0	−443.9	280.9
HCHO	g	−108.7	−102.7	218.8
CH$_3$OH	g	−201.0	−162.3	239.9
CH$_3$OH	l	−238.7	−166.3	126.7
C$_2$H$_5$OH	l	−277.6	−174.7	160.7
C$_2$H$_5$OH	g	−234.8	−167.8	281.6
CH$_3$COOH	l	−484.1	−389.3	159.8

[a] The Scientific Group Thermodata Europe (SGTE) is a consortium of seven European centers engaged in the critical assessment and compilation of thermodynamic data for inorganic and metallurgical substances of all types.

[b] g, hypothetical ideal gas at unit fugacity; aq, hypothetical ideal solution of unit molality: l, s, pure liquid or crystal at 1 bar pressure.

Answers

Chapter 1

1.1 $0 \text{ K} = -273.17^\circ\text{C}$
1.2 73.1 J
1.3 (a) Two stages compared with three
     (b) 1615 g fuel oil

Chapter 2

2.1 (a) $-56.9 \text{ kJ mol}^{-1}$
     (b) $+15.5 \text{ kJ mol}^{-1}$
2.2 7.33 kJ mol^{-1}
2.3 $-36.4 \text{ kJ mol}^{-1}$
2.4 $\Delta_{\text{isom}}H = 0, -5.4, -3.1, -8.1,$ and $-14.4 \text{ kJ mol}^{-1}$, respectively
2.5 17.94 kJ g^{-1}
2.6 $-145.8 \text{ kJ mol}^{-1}$
2.7 $-149.5 \text{ kJ mol}^{-1}$
2.8 $21.48 \text{ kJ mol}^{-1}$
2.9 Heats of combustion $(\text{kJ g}^{-1}) = 50.33 \; (C_3H_8)$ and $49.50 \; (C_4H_{10})$
2.10 $-718.2 \text{ kJ mol}^{-1}$

Chapter 3

3.1 $w = 0, q_v = \Delta U = -4289 \text{ kJ mol}^{-1}, \Delta H = -4293 \text{ kJ mol}^{-1}$
3.2 $\Delta_c H^0 = -694.7 \text{ kJ mol}^{-1}, \Delta_f H^0 = -1502.5 \text{ kJ mol}^{-1}$
3.3 $-92.93 \text{ kJ mol}^{-1}$
3.4 $\Delta C_P = 33.1 \text{ J mol}^{-1} \text{ K}^{-1}, \Delta H^0_{1500} = 425.4 \text{ kJ mol}^{-1}$
3.5 $\Delta C_P = 67.58 - (0.6658 \times 10^{-2}T) - (8.155 \times 10^{-5}T^2) + 3.631 \times 10^{-8}T^3$
     $\Delta H^0_{1500^\circ\text{C}} = 405.0 \text{ kJ mol}^{-1}$
3.6 (a) $17.90 \text{ kJ mol}^{-1}$
     (b) $2.95 \times 10^3 \text{ kg s}^{-1}$
3.7 $58.98 \text{ kJ mol}^{-1}$
3.8 7266 K
3.9 Cubic equation is:
     $-\Delta H^0 = 757,000 = 112.5T + 2.104 \times 10^{-2}T^2 - 2.015 \times 10^{-6}T^3 - 35,371$
     $T = 4744 \text{ K}$
3.10 $-122.2 \text{ kJ mol}^{-1}$
3.11 49.3 kJ mol^{-1}

Chapter 4

4.1 (a), (b) $CH_3CH_2CH_2CH_3$ (more flexibility)
     (c) N_2O_4 (more atoms)
     (d) C_3H_8 (more atoms)
4.2 $\Delta S^0 = 220.2$ J mol^{-1} K^{-1}
4.3 224.7 J mol^{-1} K^{-1}
4.4 $\Delta H^0 = 49.98$ kJ mol^{-1}, $\Delta S^0 = 336.3$ J mol^{-1} K^{-1}
4.5 $\Delta H^0 = 133.0$ kJ mol^{-1}, $\Delta S^0 = 531.5$ J mol^{-1} K^{-1}

Chapter 5

5.1 $\Delta S^0 = -233.1, +82.9, +175.9, -4.4,$ and -145.5 J mol^{-1} K^{-1}, respectively
5.2 2.68×10^4 J K^{-1}
5.3 -35.1 J K^{-1}
5.4 $T = 301.4$ K, $\Delta S_{ball} = -65.8$ J K^{-1}, $\Delta S_{water} = +87.0$ J K^{-1}, ball water, $\Delta S_{tot} = +21.2$ J K^{-1}
5.5 21.1 J mol^{-1} K^{-1}
5.6 37.2 J mol^{-1} K^{-1}
5.7 20.2 J mol^{-1} K^{-1}
5.8 22.1 (WF_6), 30.2 (OsF_6), 26.0 (IrF_6) J mol^{-1} K^{-1}
5.9 $\Delta S^0_{25°C} = 160.2$ J mol^{-1} K^{-1}, $\Delta S^0_{600°C} = 158.1$ J mol^{-1} K^{-1}

Chapter 6

6.1 $\Delta S = 26.8$ J K^{-1}, $\Delta G = -8.75$ kJ.
6.2 $SnCl_4$ 376 (387), cyclohexane 342 (354), acetone 330 (329), ethane 181 (184.9) K (actual values in parentheses).
6.3 (a) $\Delta G^0_{700} = -603.4$ kJ mol^{-1}.
     (b) $T = 1494$ K.
6.4 $\Delta S^0 = -201.2$ kJ mol^{-1}. Reaction would occur even at reduced activity of Na_2O in glass.
6.5 $\Delta G^0_{1100} = +5.35$, $\Delta G^0_{1200} = -3.99$ kJ mol^{-1}. PdO loses oxygen above 1158 K.
6.6 -112.04 kJ mol^{-1}.
6.7 (a) 211.0 J mol^{-1} K^{-1}.
     (b) Back reaction occurs at lower temperatures.
     (c) No net bonds are made or broken; $\Delta_f H^0(Si(g))$ is 446 kJ mol^{-1}, so ΔH is reasonable.
6.8 Assuming ΔH constant, $\Delta S = 162.8$ J mol^{-1} K^{-1}, positive ΔS indicates high-temperature occurrence.

Chapter 7

7.1 (a) 242.7 kJ mol^{-1} SiH_4.
     (b) -24.21 kJ mol^{-1} SiH_4.
7.2 (a) $\Delta G^0_I = -31.53$, $\Delta G^0_{II} = -55.04$ kJ.
     (b) -23.51 kJ mol^{-1} SiO_2.

7.3   (a) $\Delta G^0 = -113.4$ kJ, $K = 2.6 \times 10^6$.

     (b) $\Delta G^0 = -577.0$ kJ, $K = 7.2 \times 10^{22}$.

     (c) $\Delta G^0 = -14.35$ kJ, $K = 2.65$.

7.4 $K = 6.63$, $\Delta G^0 = -16.81$ kJ mol^{-1} CO_2.

7.5 $K = 0.1419$, fraction dissociated $= 0.185$.

7.6 $\Delta G^0 = -383.5$ kJ mol^{-1} from reaction isotherm.

7.7 $\Delta_{vap}G = +9.39$ kJ mol^{-1} H_2O, specified process has $\Delta G = 12.304$ kJ mol^{-1}, giving $P = 6.2 \times 10^{-3}$ bar $= 4.64$ mm Hg.

7.8 1.2×10^{-11} bar.

7.9   26.8 kJ mol^{-1}.

     $\Delta G^0_{298} = -25.1$ kJ mol^{-1} is promisingly negative, but rates are very slow.

7.10 $\Delta G^0 = -1.893$ kJ, $K = 1.42$.

7.11 (a) $G(\xi) = (2 - 2\xi)\left[-125{,}611 + 5405 \ln \dfrac{2 - 2\xi}{2 + \xi} \right]$

$$+ 2\xi \left[-51{,}336 + 5405 \ln \dfrac{2\xi}{2 + \xi} \right] + \xi \left[-150{,}446 + 5405 \ln \dfrac{\xi}{2 + \xi} \right].$$

(b)

(c) With the help of Mathcad:

$$\left(\dfrac{\partial G}{\partial \xi} \right) = 5405 \ln \dfrac{\xi}{\xi + 2} - 10{,}810 \ln \dfrac{2\xi - 2}{\xi + 2} + 10{,}810 \ln \dfrac{2\xi}{\xi + 2}$$

$$- 5405 \left[\dfrac{\xi}{(\xi + 2)^2} - \dfrac{1}{\xi + 2} \right](\xi + 2) + 5405 \left[\dfrac{2}{\xi + 2} - \dfrac{2\xi}{(\xi + 2)^2} \right](\xi + 2)$$

$$- \dfrac{(\xi + 2)(10{,}810\xi - 10{,}810 \left[\dfrac{2}{\xi + 2} - \dfrac{2\xi - 2}{(\xi + 2)^2} \right]}{2\xi - 2} - 1896.$$

(d) $\xi_{eq} = 0.70$ and $K_p = 1.42$.

7.12 (a) $G(\xi) = -153{,}770(1-\xi) + (1-\xi)\{-260{,}796 + 14{,}134 \ln(1-\xi)\}$

$+ \xi\{-396{,}309 + 14{,}134 \ln\xi\}$.

(b)

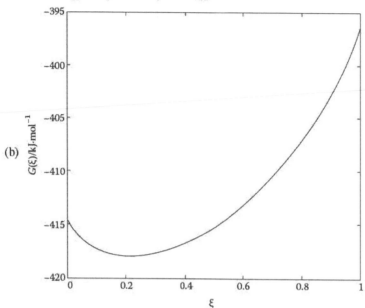

(c) $\left(\dfrac{\partial G}{\partial \xi}\right)_{T,P} = 14{,}134 \ln\xi - 14{,}134 \ln(1-\xi) + 18{,}257$.

(d) $\xi_{eq} = 0.216$, $K_p = 0.28$.

(e) Results are the same as in Example 7.4, except that the graph is shifted up +531.1 kJ on the y-axis.

Chapter 8

8.1 $\Delta_{vap}H = 40.8$ kJ mol^{-1}, $\Delta_{vap}S = 121.3$ J mol^{-1} K^{-1}

8.2 $\ln p(\text{bar}) = -5141/T + 13.79$, 361.4 K

8.3 $\Delta_{vap}H = 67.5$ kJ mol^{-1}, $\Delta_{sub}H = 128.0$ kJ mol^{-1}, $T_M = 393$ K

8.4 $\Delta_{dis}H^0 = 195.8$ kJ mol^{-1}, $\Delta_f H^0$ (Br(g)) = 112.7 kJ mol^{-1}

8.5 For $CO_2 + C \to 2CO$, $K_{p,1073} = 7.02$, $K_{p,\,1173} = 36.24$, $\Delta H^0 = 171.8$ kJ mol^{-1}; For $2CO_2 \to 2CO + O_2$, $\Delta G^0 = 361.2$ kJ mol^{-1}, $\Delta H^0 = 566.8$ kJ mol^{-1}, and $\Delta S^0 = 175.3$ J mol^{-1} K^{-1} at 1173 K

8.6 $\Delta G^0_{524} = -0.71$ kJ mol^{-1}, $\Delta G^0_{580} = -9.76$ kJ mol^{-1}, $\Delta H^0 = 84.01$ kJ mol^{-1}

8.7 $\Delta G^0_T = -33.81 + 6.63 \times 10^{-3}T$ in kJ mol^{-1}, $\Delta_r G^0_{1120^\circ C} = 323.4$ kJ mol^{-1} O_2, $P_{O_2} = 3.1 \times 10^{-12}$ bar

8.8 $\Delta_f H^0_{2000} = -248.1$ kJ mol^{-1}, $DH^0(\text{B–N}) = 643$ kJ mol^{-1}

8.9 $\Delta_f H^0_{298}$ for neutralization = -56.6 kJ mol^{-1}
Using $\Delta_f H^0$ values for H$^+$ + OH$^-$ $\to$ H$_2$O(l), $\Delta_f H^0$(OH$^-$) = -229.2 kJ mol^{-1}

8.10 0.224 bar
8.11 115.44°C

Chapter 9

9.1 3.2×10^{-28}
9.2 (a) $Fe + 2OH^- \rightarrow FeO + H_2O + 2e$
   $Ni_2O_3 + H_2O + 2e \rightarrow 2NiO + 2OH^-$
   (b) $\Delta G(\text{cell}) = -245.1$ kJ mol^{-1} Fe
   $\Delta_f G^0(Ni_2O_3) = -431.1$ kJ mol^{-1}
9.3 -0.711 V
9.4 (c) $E^0 = +0.0528$ V
9.5 $E = +0.0207$ V
9.6 96,106 V coulomb mol^{-1}
9.7 -0.0203 V
9.8 $S^0_{298}(ReO_3) = 80.8$ J mol^{-1} K^{-1}
9.9 -65.6 J mol^{-1} K^{-1}, cell value, for Appendix III $\Delta S^0 = -62.5$ J mol^{-1} K^{-1}
9.10 (a) $Ru + 2O^{2-} \rightarrow RuO_2 + 4e$
   $O_2 + 4e \rightarrow 2O^{2-}$
   (b) 135.2, -130.8, -126.5, and -131.6 kJ mol^{-1}
   (c) $\Delta_f G^0(RuO_2) = -306.8 + 0.174T$
   (d) $\Delta_f H^0(RuO_2) = -306.8$ kJ mol^{-1}, $\Delta_r S^0 = -174$ J mol^{-1} K^{-1}
9.11 $\Delta G = -20.05$, $\Delta H = -25.46$ kJ mol^{-1} Ag,
   $\Delta S = -18.1$ J mol^{-1} K^{-1}, solid components not pure crystals, activities non-zero
9.12 (a) 30.9 J mol^{-1} K^{-1}
   (b) 8.11 kJ mol^{-1}
9.13 $\Delta G = -98.28$ kJ mol^{-1} Hg, $\Delta S = -3.9$ J mol^{-1} K^{-1}
9.14 (a) 1.75×10^{-4}
   (b) 9.38×10^{-16}
   (c) 3.52×10^{-9}
9.15 (a) + : $AgCl(s) + e \rightarrow Ag(s) + Cl^-(aq)$
   − : $Ag(s) \rightarrow Ag^+(aq) + e$
   (b) -0.577 V
   (c) 1.76×10^{-10}
   (d) 1.80×10^{-10}

Chapter 10

10.1 $\Delta G^0_{623.15} = 46.1$ kJ mol^{-1} CO, $K = 1.37 \times 10^{-4}$.
10.2 (a) 4.86×10^{-5}.
   (b) $p(CH_3OH) = 52.6$ bar, % conversion CO = 44.4.
   (c) $p(CH_3OH) = 82.4$, $p(CO) = 55.9$, $p(H_2) = 111.7$ bar, % conversion CO = 59.6.
10.3 Steel oxidizes after Zr, before Mo.
10.4 UO_2 does not oxidize stainless steel. UO_2 is more stable than the oxide of stainless steel over the whole temperature range.

10.5 $\Delta G^0 = 55$ kJ/2 mol Mg, $\Delta G = -40.8$ kJ/2 mol Mg.
10.6 3.31 V.
10.7 1680 K.
10.8 $\Delta G^0 \approx -64.0$ kJ mol^{-1}, $K = 423$.
10.9 $E^0 = 2.50$ V; 7.53 V actually used.

Suggested Further Reading

Three introductory texts on approximately the same level as this book are:

Cardew, M. H. 1994. *Thermodynamics for Chemists and Chemical Engineers: A User-Friendly Introduction.* Ellis Horwood Series in Chemical Engineering. Englewood Cliffs, NJ: Prentice-Hall.

Price, G. 1998. *Thermodynamics of Chemical Processes.* New York, NY: Oxford University Press.

Smith, E. B. 2004. *Basic Chemical Thermodynamics.* 5th ed. New York, NY: Oxford University Press.

The following more advanced textbooks are recommended for their clarity:

Atkins, P. W. 2006. *Physical Chemistry.* 8th ed. New York, NY: Oxford University Press.

Caldin, E. F. 1958. *An introduction to Chemical Thermodynamics.* New York, NY: Oxford University Press.*

Çengel, Y. A., and M. A. Boles. 2002. *Thermodynamics An Engineering Approach.* 4th ed. New York, NY: McGraw-Hill.

Denbigh, K. 1997. *The Principles of Chemical Equilibrium.* West Nyack, NY: Cambridge University Press.

Guggenheim, E. A. 1968. *Elements of Chemical Thermodynamics.* Monographs for Teachers, 12. London: Royal Institute of Chemistry.*

Klotz, I. M., and R. M. Rosenberg. 2008. *Chemical Thermodynamics: Basic Theory and Methods.* 7th ed. New York, NY: John Wiley and Sons.

Kondepudi, D. 2008. *Introduction to Modern Thermodynamics.* 1st ed. New York, NY: John Wiley and Sons.

Kyle, B. G. 1999. *Chemical and Process Thermodynamics.* 3rd ed. Englewood Cliffs, NJ: Prentice-Hall.

Pitzer, K. S. 1995. *Thermodynamics.* 3rd ed. New York, NY: McGraw-Hill.*

McQuarrie, D. A., and J. D. Simon. 1999. *Molecular Thermodynamics.* New York, NY: University Science Books.

Reid, C. E. 1990. *Chemical Thermodynamics.* New York, NY: McGraw-Hill.*

Other more specific references apply to some sections of this book:

Treptow, R. S. 1996. Free energy versus extent of reaction. *J Chem Ed* 73:51–54. (Free Energy versus Extent of Reaction, Chapter 7)

Austin, L. G. 1959. Fuel cells. *Sci Am* 201:72–78. (Fuel cells, Chapter 9)

Ball, C. J. P. 1956. The history of magnesium. *J Inst Metals* 84:399–411. (Magnesium, Chapter 10)

Carle, T. C., and D. M. Stewart. 1962. Synthetic ethanol production. *Chem Ind* 830–839. (Ethanol, Chapter 10)

Dehoff, R. T. 1993. *Thermodynamics in Material Science.* McGraw-Hill, 322–332. (Ellingham Diagrams, Chapter 10)

Glassman, I. 1965. The chemistry of propellants. *Am Sci* 53:508–524. (Propellants, Chapter 2)

Hack, K. 2008. *The SGTE Casebook: Thermodynamics at Work.* 2nd ed. Boca Raton, FL: CRC Press, 415. (Silicon, Chapter 11)

---

* This book is out of print but is probably available at www.amazon.com.

Index